BATTLE OF WITS

By the same author

ALFA-ROMEO: A HISTORY
CARS OF THE TARGA FLORIO

BATTLE OF WITS

A History of Psychology & Deception in Modern Warfare

BY

DAVID OWEN

LONDON
LEO COOPER

First published in Great Britain 1978 by
LEO COOPER LTD.
196 Shaftesbury Avenue, London WC2H 8JL

Copyright © 1978 David Owen
ISBN 0 85052 217 X

Set in 11pt. Imprint type and
Printed in Great Britain by
Ebenezer Baylis and Son Ltd.
The Trinity Press, Worcester, and London

Contents

INTRODUCTION

NAPOLEON ASSESSED the value of the moral factors in warfare as three times that of the material. And in spite of every advance in training, in weaponry, in strategy or in tactics, the pyschological factors—morale, spirit, initiative, surprise, secrecy and deception—are still paramount, be it in urban Belfast or in the jungles of South-East Asia, just as they were on the coastal plain of Marathon more than two thousand years ago.

Man is above all a thinking animal. And, as the only animal who fights his own kind to the death in mortal combat, all but the least imaginative of fighting men find themselves locked in a deep inner conflict. On the one hand, rational thought processes tell soldiers they must adopt a given course of action to defeat the enemy, and to avoid the defeat of their own side. The motives may vary—from hatred of whatever the enemy represents to love of one's own freedom, from fear for one's family to the dictates of discipline, tradition and self-respect, or the weight of clear and unmistakable orders. All these pressures tug the soldier in one direction; but pulling him the opposite way is the deeper instinct of self-preservation. Standing there to be shot at may achieve all the right objectives in time—but running and hiding may seem more likely to preserve his own irreplaceable life in the immediate short term.

Which of these two arguments triumphs depends on a host of factors, quite apart from his own mental make-up. His hopes of victory, the threat of defeat, his own religious or political motivation, his faith in his leaders, his training and experience, the penalties for failure and the behaviour of those around him—all have their effect in tilting the balance to one side or the other. Even subtle factors like regimental tradition, battle honours, education and propaganda, not to mention individual worries about family or finances, can at times be decisive.

Very rarely, throughout the whole history of warfare, does one side triumph over the other by killing, or even wounding, all the enemy. Victory and defeat come much earlier, when one side feel they have no chance of gaining their own objective—either winning on their own account, or some lesser aim such as delaying their enemy sufficiently to gain time for their own side, or the simple maintenance of their own regimental traditions and reputation. Once the moment arrives when they feel that continued resistance is pointless, then, whatever the relative numbers or firepower, the issue is decided. Enough individuals revert to self-preservation, and resistance crumbles, often with bewildering speed.

Obviously the tougher and more experienced the troops and the more competent the commanders, the harder it will be to push them to this point. But fighting spirit, even among seasoned troops, can be a fragile quality. In most armies, or in individual units, a casualty rate of thirty per cent will be enough to produce a total psychological collapse. Recovery may be equally swift, especially with rest, reinforcement and re-equipment, but, temporarily at least, even the bravest and most formidable units are rendered incapable of fighting.

Seen in this light, the development of new tactics and weapons can be regarded as a series of attempts to bring the enemy to this point as quickly, as certainly and as cheaply, in terms of one's own losses, as possible. With superior weapons, greater numbers, better use of ground, more firepower and faster manœuvres, this desirable state of affairs can be reached relatively quickly and easily. Yet, throughout history, examples abound where victory was gained above all by a direct attack on the minds of the enemy—either as part of, or even instead of, a direct physical attack—by frightening them, misleading them, surprising them, outguessing them or demoralizing them. So powerful are these factors that all too often they can make up for inferior weapons, smaller numbers, lesser experience and all manner of other material disadvantages. Skilfully used, they have turned the tables completely, upsetting all the so-called rules and precedents of the art of war.

This kind of attack on the enemy's morale is psychological warfare in its purest sense. But applied psychology—psychology used to create a particular idea in the mind of the enemy—can open up a much wider range of equally devastating effects. If the enemy can be convinced that one's own forces are more numerous than they really are, if he can be persuaded that one's main attack will come at a particular spot and a particular time so definitely that he shifts his own forces to counter the expected blow, this will leave him far more vulnerable to a real attack delivered at a different time at another place.

Unfortunately, there are all kinds of pitfalls on the road to this objective. Enemies, in the real world, are rarely fools. Quite apart from the danger of their mounting a deception of their own, there is always the possibility that they will react in the wrong way. Cases abound in history of the enemy reaching the wrong conclusions. He may see through the deception drawn up to fool him and only pretend to believe it. He may refuse to see information planted for his benefit—or he may refuse to recognize it as genuine. He may, for reasons of his own, decide coincidentally that one's own real plans are the most likely course of action, whatever the evidence to the contrary. He may, finally, actually be a fool—foolish enough to hit on the truth after missing the most obvious and well-planned deception.

The classic information and intelligence dilemma is inescapable in this kind of battle. Suppose one's own forces are about to attack at point A, but the enemy needs to be convinced that the attack will come at point B. Some subtle hints that the attack will arrive at B are clearly needed, but the balance is delicate, and absolutely crucial: too few hints, or too much subtlety, and the enemy may either miss them or discount them, deciding that an attack at A is more likely because he has no real information to the contrary. On the other hand, should the planted information, the calculated leaks and the deliberately uncovered clues, be slightly too obvious, then he may guess the truth. Once he decides that the indicated attack at B is a deception, he will conclude that the real blow will fall at A.

There are all kinds of subtleties which can be brought to bear in such a psychological struggle—bluffs, double-bluffs and treble-bluffs— but there are several general principles worth keeping in mind. For example, the more nearly a deception coincides with the enemy's own preconceived assessment of the situation, the more likely it is that he will be fooled. Likewise, the more indirect the information fed to him in order to create a particular picture in his mind, and the more work he has to do for himself before reaching the conclusions he is intended to draw, then the more likely he is to accept the result as the truth.

The same essential conflict applies to information received about one's enemy. Is a given clue a genuine indication of a coming attack, or is it part of the enemy's own deceptions? If a particular course of action seems totally obvious, then this might be a deliberate and not very careful deception—or it might be the truth. Very often, the choice of a particular deception will depend on the facts of military life which cannot be hidden from the enemy's eyes. If he can easily see particular signs of troop convoys, ship movements, defence works and so on, then the chosen deception must fit all these factors. Anything which can be seen which does not fit the picture will tend to reveal it as a lie.

Another possibility is the multi-layered deception. Assuming that an alert enemy is wide awake to the possibility of deception, then he will tend to be over-suspicious of the first pieces of information fed to him. If, on the other hand, further evidence emerges to discredit it, then he will be more disposed to accept the new picture as the truth. In some cases it may take three or more layers before the final picture is taken as the real situation, but if the enemy can then be fed enough indirect and convincing information for him to feel he has ferreted out the ultimate reality for himself, in many cases he will cling to this picture in spite of any new information to the contrary.

So how can facts, or apparent facts, be assessed? In general, the more widespread and indirect the sources of information which point towards a particular picture, the more likely it is to be the truth. Yet even here the same dilemma is inescapable, since an alert enemy will know that only by counterfeiting information of this nature does he stand any real chance of mounting a successful deception. And although ideas and tactics like these play their full part in applied psychological warfare, directed towards a particular deception plan, they have their value in direct psychological attack too. If, for example, propaganda is intended to worry the enemy and demoralize him on one particular issue, then the attack will usually be much more successful if the facts revealed fit what the enemy already knows to be true—and, equally important, if the inference to be drawn from the propaganda is put in such a way that the recipient feels he has only reached the right conclusion by reading between the lines of the message for himself. Any propaganda will also carry much more psychological weight if it seems to come from one's own side: especially if the printing, the language, the content and point of view all reinforce this impression.

In one sense the basic precepts of psychological warfare are as old as war itself. In the ancient world, defenders of besieged fortresses were frightened by psychological weapons such as Greek Fire, or depressed by the threat of pestilence from dead horses and corpses of prisoners hurled over the ramparts by catapults. Alexander the Great, knowing how much the fighting spirit of an Eastern army depended on its ruler and commander-in-chief, framed his tactics to kill or capture the enemy king as quickly and economically as possible.

The formidable Roman military machine built up a psychological ascendancy over their enemies by training and routine. Whenever the legionary was not fighting, he was practising the business of fighting, through sound practical training designed to knock any lingering ideas of the romance and excitement of war out of the heads of young recruits. Practice weapons were twice as heavy as those used in action. After every day's march, the troops would build a properly fortified camp before retiring for the night, as a matter of routine. The layout was entirely standardized so that the legion would form ranks instantly when the alarm was sounded, even in pitch darkness. Discipline was harsh, the idea being to turn the anger of the soldiers outwards against the enemy, and even the most meagre concessions were valued as cherished privileges by the long-suffering troops.

Later civilizations realized the benefits of psychology. The Byzantines anticipated Napoleon by organizing units of non-standard size to avoid revealing the real strength of their forces to an observant enemy. They planned campaigns carefully, attacking the Scythians in March when the horses on which the enemy depended would be weak from lack of winter forage. They attacked marsh-dwelling enemies when the rivers were frozen, making manœuvre easy and robbing them of semi-submerged ambush tactics. Mountain-dwelling opponents were attacked in the snow, when their tracks were plainly revealed, while adversaries in the warmer Mediterranean countries were attacked in cold and rainy weather when their spirits were at their lowest.

The Chinese used inventions like fire-crackers, rockets and primitive flame-throwers to strike terror into their enemies rather than for any real advantages in firepower. But it was China's conqueror, Genghis Khan, who refined terror into a truly deadly weapon—cities which surrendered at his advance were ruthlessly sacked, and many of their inhabitants murdered. Their only doubtful consolation was that any city which *did* resist was afterwards burned to the ground, with the massacre of *all* the surviving population. On the other hand, this stark choice was obscured by a sophisticated network of Fifth Columnists which moved in advance of the Mongol army, spreading rumours that *this* time any city which willingly opened its gates to the invaders would be spared.

In its most sophisticated form, however, the use of psychology as a weapon of war belongs to the twentieth century. Two hundred years and more ago, the points of contact between opposing forces were close, direct and simple. If one's enemy was in sight, then there was a limit to the deceptions which could be brought into action. On the other hand, once the enemy was out of sight, then all contact was lost in most cases—and once contact was lost, the flow of information which could confuse, demoralize or defeat the other side was cut off with it.

But, from the First World War onwards, the entire character of warfare changed. From now on, armies and commands remained in contact with one another for years on end. They could watch each other's positions, eavesdrop on one another's movements and intercept their opponents' signals as never before. Gone was much of the fog of war; now adversaries watched each other as closely and as carefully as poker players, and the opportunities to deceive and discourage became greater than ever.

Along with new opportunities came new challenges. Feeding the enemy information or propaganda may have become easier—but it was also much easier for him to spot any flaws or inconsistencies in what he saw. Every new advance in technology, in weaponry and in tactics made the achievement of a psychological coup both more difficult and yet more worthwhile—for as the problems grew, so did the rewards.

This is the story which this book sets out to tell. In the main, it is a story of British ingenuity pitted against the wits of German adversaries. There are good reasons for this concentration. On the Allied side, in both World Wars, the British remained the leaders and the principal innovators in this specialized area of warfare, though later our American allies were to adopt and develop well-established tactics. On the Axis side, the Germans were the masters of propaganda as well as the most formidable opponents. The Japanese, on the other hand, whatever their undeniable military qualities, were less well versed in this particular field. At the same time, the conditions of warfare in the Far East were relatively so remote and primitive that the opportunities for psychological fighting were severely limited.

Much of the background to the story will be familiar, in terms of the battles and campaigns which appear in its pages. Others will be less well known, or conspicuous by their absence. What will be different is that for the first time the influence of psychological factors on military history is told as a single coherent narrative, showing how great an effect these vital but still largely ignored factors have exerted on modern warfare. As a view of the working of human nature and human ingenuity, the story is an absorbing one—and an important one too. For as men themselves, not to mention their ways of waging war, grow more complex and more sophisticated, so these methods of psychological attack and defence become ever more vital. Now, when the firepower at the disposal of a single soldier often exceeds the hitting power of an entire division in the Second World War, the importance of the psychological factors in war becomes paramount. For as warfare itself grows ever more deadly in its implications for the future of the world as a whole, then confusing, outwitting, discouraging and defeating one's enemy by deceiving him and persuading him that his cause is lost before battle begins become infinitely more desirable than simply trying to kill him.

CHAPTER ONE

THE WAR TO END WARS

THE STORY begins at the start of the Great War. The first clashes of twentieth-century mechanized warfare came as an appalling shock, even to highly-trained professional armies, for the ultimate ruler of the battlefield turned out to be neither the cavalry sabre nor the infantry rifle but the ugly, unromantic yet undeniably efficient machine-gun. By the time the Allies had come to terms with the new kind of war, where all the dash and courage in the world counted for nothing against massed firepower, it was almost too late. For the time being, all the advantages—psychological and material—lay with the defence, so both sides went firmly on to the defensive. Since men could only survive by burrowing into the ground for safety, as the Americans had found during their Civil War, what often began as a temporary local expedient was soon extended and developed into a permanent line of fortifications. As both armies strove to turn one another's flanks, a digging race began. A steel-bound corset of trenches and barbed wire began to extend southwards towards the Swiss frontier and northwards to the Channel coast.

By this time, however, the Germans were in the strongest position. Their headlong rush across the French frontier which had opened the war may have been stopped short of Paris and a quick victory, but as the Western Front stiffened into deadlock they were ideally placed to win a war of attrition. They held nearly all Belgium as well as that part of northern France which contained much of the country's vital industries and coal reserves.

But these advantages had their own disadvantages, for with the gains came the problem of holding on to them. Predictably, German control over Belgium roused in most inhabitants anger and hatred—and the possibility of guerrilla action was always there. In particular the Germans felt vulnerable to civilian bands of snipers, such as they had met in 1870, and any sign of defiance on the part of the conquered population was crushed as a matter of deliberate policy for two reasons: to show their own troops that any trouble from trigger-happy civilians

1

would be ruthlessly put down, and secondly to cow their new subjects into total obedience. Civilians caught in possession of arms were shot, as were the inhabitants of villages where snipers had killed German soldiers. Hostages were taken to ensure the good behaviour of the populace, and these too were shot at any sign of unrest, while villages were burned as examples to others of the folly of resistance.

This behaviour had a solid enough pedigree. Clausewitz himself had advocated the use of terror as a means of producing a quick and decisive decision instead of a long-drawn-out battle of attrition. When the Belgians fought back, blowing up bridges and cutting railway lines, the German use of terror escalated beyond all bounds. More than 200 civilians in the village of Andenne were shot on 20 and 21 August, 1914. A day later the town of Tamines was looted and 384 people—men, women and children—shot and bayoneted by firing squads. In the city of Namur the Germans took ten hostages from every street as a guarantee of good behaviour—any activity by snipers and the hostages were shot. Soon this had escalated to a hostage from every household of every town or village in which German troops were billeted. On 23 August, 612 people, including a three-week-old baby, were executed in Dinant by two firing squads, as punishment for 'hampering the reconstruction of bridges', and the city was looted and burned. But the atrocity which most aroused the anger of the world was the sack of Louvain, when for almost a week the city was the scene of burning, shooting and destruction, culminating in the total loss of the University and the unique medieval library—all as a reprisal for an attack by Belgian troops on German units outside the city.

As a result the tone of the propaganda war which both sides were by now conducting against one another underwent a sharp change. Hitherto British newspapers had been content to laugh at the Germans. But as reports of German atrocities began to filter through to the outside world, the attitude began to harden. Now the Germans were seen as sadistic brutes, delighting in causing suffering for its own sake. Always eager for sensation, the press fed its own hunger for atrocity stories by embroidering any promising rumours which came to hand, and finally by outright invention. German soldiers using children as living shields, mutilating young women, shooting helpless nuns and bayoneting babies for sport became common currency, fanning a new wave of public anger and hatred of the enemy.

Not that the Germans themselves were above a little invention for propaganda purposes. They felt that a demonstration like the burning of Louvain would act as a salutary reminder of German strength—but, nevertheless, some justification was needed for home consumption to explain the harsh treatment meted out to the Belgians. Lurid tales of armed bands of snipers, led by priests and capable of every kind of atrocity against innocent German soldiers, filled the German papers. Belgian women and children were known to have cut the throats or gouged out the eyes of German wounded waiting for help. Again public anger was roused, and again the gulf of hatred between the warring populations widened and deepened.

Fanning the flames of hatred for the Germans was one thing; defeating them was more difficult. Time and again head-on attacks failed with terrible losses. But because the front stretched unbroken from sea to mountains, no outflanking movements were possible. Yet, time and again the psychological advantages of surprise and deception were wantonly discarded. To the staffs who planned the attacks little mattered beside the weight of shells in the artillery barrages and the masses of men deployed in ever more frantic efforts to break through the enemy lines.

The Somme was a classic example of the British commanders' scant regard for the value of out-thinking the enemy as a preparation to outfighting him. As soon as they took over from the French, they began the usual policy of 'keeping the enemy on his toes' by constant raids and sniper fire. This had two results: firstly, casualties on the British side climbed quickly (over the whole British front, 125,000 men were lost in raids in just six months); secondly the Germans were alerted by the increased activity. Their defences had tended to fall into disuse under the quiet live-and-let-live policy of the French in this area; now they drafted in thousands of Russian prisoners and put them to work building a formidable defence position four lines deep, so that when the attack came, the task facing the British was that much more difficult.

Next came the preparations, on a larger scale than for any previous attack. Yet, although the Germans overlooked the British positions, no attempt was made to keep the build-up concealed, nor was there any attempt made to deceive the enemy as to where the blow would be aimed. The incredulous Germans watched the British digging trenches, laying railway tracks, repairing roads and cramming the whole Somme area with troops, guns and supplies. So blindingly obvious was the whole process that the Germans decided at first that it was impossible for the British to be considering an attack in this area. The whole operation must be an elaborate deception, a double bluff, designed to distract their attention from another, more vulnerable part of the line.

The truth was not long in dawning. The Germans, like the British, were now carrying out regular trench raids to capture prisoners and extract information. But whereas the costly British sorties failed to penetrate far enough into the German defences to discover anything about their formidable strength, the Germans were able to find out almost everything they wanted about the new formations being brought in ready for the attack. As each unit took over its section of front line, it would be greeted by catcalls and shouts. The Royal Welch Fusiliers were cursed as 'Bloody Welsh murderers', the Royal Sussex Regiment as 'Bastard Sussexers', and other units would be asked questions about their home towns which showed everyone just how much the Germans already knew.

Then the leaks began. Speeches at home, remarks over dinner tables in neutral capitals, information from prisoners and reports in French and neutral papers, all gave pointers to the coming attack. Belatedly, the British High Command set in train a half-hearted deception plan, digging trenches further to the north of the attack area, but by now the Germans were convinced they knew the truth and set to work reinforcing their defences, with hidden machine-gun emplacements concealed in the hillsides, with shell-proof dugouts forty feet under ground, with sleeping quarters, hospitals and supply and ammunition stores supplied by railways.

Yet the Germans knew full well the advantages of lulling the enemy into a state of false security. For all this furious activity, nothing could be seen by the Allied aircraft which ranged all over the front during the hours of daylight. On the other hand, German aircraft watched the training exercises and the progress of the British troops marching between lines of tape laid out to represent the trench systems—and thereby built up a very accurate picture of the size, weight and sequence of the attack plan. When the initial diversionary attack at Gommecourt was being practised, the Germans were able to count the divisions which would actually be involved by adding up the number of observation balloons watching the exercise. The answer told them clearly that the main blow would come further south, and they finished their preparations accordingly. But so complacent were the British commanders, and so sure that the Germans would not notice what was happening, that the Chief Intelligence Officer on Haig's staff had interpreted the movements of German units as routine rest and retraining reshuffles. As late as 28 June, when the preliminary bombardment had been pounding the German positions for four whole days, he still considered the enemy had no idea an attack in force was on the way.

The attack was a catastrophe. Although the heavy guns shelled the

German positions for six days the thick barbed-wire entanglements were not cut up sufficiently for the attackers to get through. The Germans had hidden their own gun positions so that they could not be pin-pointed by British observers, and few of them were knocked out by the bombardment. The infantry themselves were weighed down with 66 lbs of stores and equipment per man. Twenty thousand died on the first day of the battle, with twice that number wounded, for gains which were pitifully disappointing.

But this was merely the beginning. For the Staff's misreading of the German resources and morale told them that another attack, though costly, might be enough to achieve the breakthrough which would make everything worthwhile. So attack succeeded attack, and although gains were made, slowly and painfully, the breakthrough never came. Casualties went on climbing and the morale of the army declined. The Germans' new fighters began to win back control of the air from the Royal Flying Corps, and the army's trump card, the new tank, was wasted by being used in small numbers in a purely local action.

This should have been a shattering blow to the Germans. Indeed it was, for the first soldiers to meet the clanking steel monsters of 1916 turned and fled in confusion. Enough tanks, used aggressively, to capitalize on their psychological impact, could have won a decisive battle. But the High Command insisted on throwing them into the Somme battle when only forty-nine were ready—too few for worthwhile gains, yet enough for the Germans to realize what was in the offing, and for them to develop anti-tank tactics. Men were briefed on the new armoured vehicles; their low speed of just over walking pace was emphasized and troops were trained to destroy them. Although tanks were to remain a formidable weapon, never again would they have the advantage of terror and surprise which their first appearance provided.

Finally, even the weather turned against the British. Although the Germans had been pushed back, the ground had been so badly pounded by shot and shell that the roads and drainage system of the whole area had been destroyed, and when the autumn rains began, the whole battlefield turned to liquid mud. Simply existing in conditions like these was such an effort that the fighting petered out for the winter and 1916 ended in misery and deadlock for attackers and defenders alike.

* * *

At sea, too, there had been grave setbacks for the Allies. At the very beginning of the war the German Navy brought off a classic deception coup, although not of its own making. So heavy had been the pre-war propaganda barrage, backing up the building of Germany's new fleet, that the Royal Navy took the threat of its new adversary very seriously indeed. Faced with its first real challenge in a century, the Admiralty credited its German opposite numbers with far more courage, more daring and more professional skill than they actually possessed. The result was that the British were so paralysed with caution that the Germans were given an almost free hand for offensive action in many areas.

One of these was the Mediterranean. On the very day war was declared a squadron under the command of Admiral Souchon, consisting of the battlecruiser *Goeben* and the light cruiser *Breslau*, was off the French North African coast. The British, with no less than three large battlecruisers at their disposal in the area, were terrified the Germans might attack the troop convoys bringing France's colonial forces to Europe. In fact, the Germans had no such intention. Souchon's orders were to sail eastwards to Turkey in an attempt to bring her into the war on the German side. But the British commander, Admiral Milne, fell into a classic trap. As the German ships sailed further and further east, he grew more and more cautious, convinced he was the intended victim of an elaborate plan designed to lure him away from the vulnerable convoy routes.

The Germans had problems of their own. They were running short of coal and needed to evade their pursuers long enough to rendezvous with colliers. But it was the British who smoothed the way for them. As they continued eastwards, Milne became increasingly anxious about HMS *Gloucester*, the shadowing cruiser which was reporting on the Germans' movements to the rest of the fleet. Might she not be sailing into a trap? So convinced was he that the Germans' real objectives lay to the west that he ordered *Gloucester* to drop back, and confined himself to patrolling the routes which the enemy would have to take if they retraced their course. In fact, freed from pursuit at the one time when their intentions might have been revealed, the Germans took on the coal they needed and finished their journey to Constantinople unmolested. On their arrival the ships were presented to Turkey as replacements for Turkish warships which had been under construction in British yards and had been taken over by the Royal Navy on the outbreak of war. This gesture helped to win over Turkish public opinion. Within weeks the two ships were leading a Turkish squadron to bombard Russian harbours in the Black Sea, and at the beginning of November, 1914, Turkey entered hostilities on the side of Germany.

The bitter lesson of this episode—not to suspect a cunning plot merely because the enemy acts in a way which clashes with the most obvious estimate of his intentions—was to take a great deal of learning. Throughout the war the German High Seas Fleet lay dormant behind its torpedo nets and shore batteries, tying down much larger British naval forces which would have been far better employed elsewhere. But what seemed to the bemused British to be enemy cunning was in fact the absence of any real plan at all in the Germans' minds as to how their heavy ships could be used offensively.

The U-boats, however, were another matter altogether. Whereas the British blockade of Germany and German trade depended mainly on surface warships which could intercept neutral merchantmen, inspect their cargoes and then escort them to ports for examination by Prize Courts, the Germans depended on surprise attack by torpedo. As the German blockade tightened, they were forced to use more and more indiscriminate methods, until in 1917 they embarked on unrestricted submarine warfare as the one way to force Britain to surrender. It was a deadly weapon, and it nearly succeeded, but it had a double edge. Neutral countries, finding their ships sunk by German torpedoes on the high seas, were less ready to complain at being intercepted and escorted by British warships, and the British blockade became much more effective as a result.

At first there was little defence against the submarine. The only way to detect submerged submarines was by using the primitive hydrophone to listen to the noise of their propellors, and this gave little information of any use in delivering an attack. Every possible expedient was tried, including the training of seals to give warning of submerged submarines in return for rewards of fresh fish! Unfortunately, when released into the Channel, fresh fish were so plentiful that the animals immediately forgot their duty and promptly deserted to the deeps. But in time new tactics were developed. The convoy system, whereby merchant vessels sailed in groups under naval escort, was revived from the days of the seventeenth-century Dutch Wars, and although it offered little direct hope of sinking submarines, it proved most effective. Psychologically, the merchant captains were encouraged by one another's company and by the presence of the naval escorts. Likewise the U-boats were deterred by the presence of warships dropping occasional precautionary depth-charges whenever their hydrophones told them submarines were in the vicinity.

In all, the submarine war cost more than 4,000 merchant ships, or more than eight million tons of shipping and the lives of 15,000 seamen.

But it cost Germany the war, in two very different ways. To begin with, the Americans, the most powerful of the neutral nations, were outraged by German tactics. The introduction of unrestricted submarine warfare was almost enough to tip the United States Government into a declaration of war. One more false move by Germany was needed and that was provided through the Admiralty's own code-breaking department, who succeeded in decoding the document which became world-famous as the Zimmermann Telegram.

This department already had some noteworthy successes to its credit. One of the first blows struck by the Royal Navy against Germany had been the cutting of the German transatlantic cables, which forced all communications between Germany and the Western hemisphere to be sent by radio, so that they could be intercepted and monitored by Allied listening stations. In the second month of the war, a German cruiser was wrecked in the Baltic and the body of one of her officers, still clutching the ship's code books, was picked up by the Russians. The code books were sent to London, where they provided the key to the entire set of naval dispatches and many of the German diplomatic codes.

By the beginning of 1917 the code-breakers had been working for several months on a German Foreign Office code known as 0075. Many telegrams using this code had been intercepted, and a list of the known code-word equivalents had been built up. So when a long message addressed to Germany's Ambassador to the United States was intercepted on 17 January, 1917, the experts were able to tell that it was signed by the German Foreign Minister, Arthur Zimmermann, and that it referred to the introduction of unrestricted submarine warfare on the 1st of the following month. Zimmermann knew that this action could bring America into the war. If it did, the Ambassador was asked to contact the President of Mexico secretly and propose an alliance between Germany and Mexico against the United States.

As more and more traffic in code 0075 was intercepted, the Admiralty experts were able to decode more and more of the telegram. On 3 February, 1917, America broke off diplomatic relations with Germany, but stopped short of declaring war on the U-boat issue. The Admiralty dilemma was a frustrating one: if they revealed the text of the telegram to the Americans, it would almost certainly bring the United States into the war on the Allied side; but the Germans would realize that their messages were being read and change their codes, thus cutting off the supply of intelligence. At the same time, the form of the message would show that the British had been intercepting neutral countries' communications, which might have the opposite psychological effect on American opinion from that intended. The problem seemed insoluble.

The answer was the brilliant inspiration of the Admiralty's Director of Naval Intelligence. He reasoned that the Ambassador in the United States, Bernstorff, would forward the message to the German Embassy in Mexico, as ordered, in a different code and a slightly different form. So a British agent in Mexico City visited the telegraph office and came away with a copy of the second telegram—sent from Bernstorff to Eckardt, the Ambassador in Mexico. This was in a totally different code, which had already been almost completely broken by the Admiralty experts. They decoded the message and found that it was indeed slightly different from the text of the original Zimmermann telegram and it was this version which was shown to the Americans on 22 February, 1917.

The Americans were staggered. The text as now decoded referred to generous financial support for Mexico and understanding of her aims to reconquer Texas, Arizona and New Mexico; while Mexico for her part would mediate between Germany and Japan. It was political and psychological dynamite. The British kept quiet about how the telegram had been found, telling the Americans it had been intercepted in Mexico. The Germans reached the same conclusion, so that the naval and Foreign Office codes were not suspected—and although the Mexicans, and even Ambassador Eckardt, denied any knowledge of the telegram so strongly that some Americans suspected an Allied plot, Zimmermann himself admitted authorship. In just a few short weeks American public opinion did a complete U-turn. Woodrow Wilson, who had been elected President on a platform of keeping America out of the war, asked for the backing of Congress to declare war on Germany. On 2 April, 1917, he got it. And Germany's most potent weapon had played its part in bringing about her defeat.

But for the time being, the position of the Allies was grim. Russia was out of the fighting, a casualty of revolution, while France was on her knees after the disasters of 1917. Until the Americans arrived in strength, any further attacks on the Germans in the West would have to be made by the British Army. The result was the offensive in Flanders— the third battle of Ypres, known more familiarly as Passchendaele. But once again, the carefully laid plans went wrong.

The preliminary attack on Arras and Vimy Ridge was successful, at the price of heavy casualties and delays which were to threaten the main attack. But it was the next phase of the build-up which showed

just how much could be achieved with careful planning and the maxi-
mum use of shock and surprise as offensive weapons. General Sir
Herbert Plumer, commanding the Second Army, was to attack the
Messines-Wytschaete ridge on the south-eastern side of the Ypres
Salient, drive the Germans off the high ground and secure the right
flank of the coming offensive.

In some ways this was an appallingly difficult assignment—attacking
uphill against defenders who had been in their positions for a long time.
But unknown to the Germans, Plumer's men had been burrowing
under their positions since 1915. Some of the tunnels went as far as
half a mile under no-man's-land and were deep enough to escape
damage from the constant shell-bursts on the surface. Every possible
precaution had been taken to prevent the Germans from realizing
what was afoot. When the tunnellers struck a layer of blue clay they
knew that any sign of this at the surface would give the game away
completely. Instead, every scrap had to be carried away at night and
reburied. The task of tunnelling in an area where the drainage system
had broken down under the shelling and where most of the ground was
flooded feet deep after every shower of rain, can only be guessed at,
but through two winters and one full summer the work went on.

Finally, at ten past three in the morning of 4 June, 1917, came the
moment of truth. Without a single rifle shot to break the calm of the
night, and without a second's warning, a total of nineteen mines and
almost 500 tons of explosive went up in a single shattering concussion.
Before the shattered fragments of steel and wood came down to earth
again, a furious bombardment from more than two thousand guns
rained shells on the battered German defences. Following the creeping
barrage came 80,000 infantry soldiers in a quick, stealthy attack. They
captured the ridge, and they held it, with casualties only 20 per cent of
those expected, at a time when most attacks lost many times their
predicted totals. The shaken Germans scarcely knew what had hit them.
Many were shot or captured as they emerged from their dugouts, and
for the first time it appeared that the British might be on the point of
breakthrough.

Yet, after such promising beginnings, the main attack did not begin
until the end of July, by which time the weather had broken. August,
1917, saw five times the rainfall of either of the two previous Augusts.
The low-lying terrain became flooded and the main attack bogged down
in the mud. Even when tanks were brought in to help at Cambrai in
November, the British were psychologically completely unprepared for
the opportunity. The armoured vehicles were bigger, more powerful

and more numerous than ever before—more than 470 tanks ready for action that day. Preparations had been carefully laid to disguise what was coming, and the Germans had no inkling that they were about to face a massive armoured assault. Everything was calculated to produce the maximum shock effect. A barrage from a thousand guns opened up without warning, ten minutes after the tanks and infantry had started to move forward. Each tank carried fascines—bundles of wood to drop into the German trenches to enable them to cross. Files of infantry followed behind the tanks, relying on them to force a passage through the wire and into the heart of the German positions.

The effect on the Germans was devastating. As they emerged from the battering of the artillery barrage, they were overwhelmed by the frightening spectacle of a massed tank attack, against which they were powerless. Tanks were familiar objects, but not tanks in such numbers. The Germans broke and ran, or were over-run. Terrified by the invulnerability of the armour and by the speed of the attack, the defence totally collapsed, and for one glorious day the way was wide open for the longed-for breakthrough. But psychological victories need to be exploited instantly; experienced troops, however badly knocked about, recover very quickly, and the Germans were fast re-organizing.

The tanks had done their job splendidly, but they were too unreliable mechanically to exploit the gap in the German lines; only one in four was still working at the end of the day. Now was the cavalry's hour but it was too slow moving up from the rear. By the time the first squadrons reached the front lines, the gap was being closed by German counter-attacks. After years of positional, almost static, warfare, the sudden opportunity of fighting a mobile campaign again was too large a pyschological challenge to cope with in the few hours during which it had been possible. So the war's fourth winter saw the traditional, and seemingly inevitable, stalemate re-establish itself.

Psychological warfare was to achieve its most brilliant victory so far against a different foe in a different theatre of war—against the Turks in Palestine. Here too the fighting had solidified into bloody deadlock before Gaza, but here there was at least an inland flank which could be turned, at the other end of the front near Beersheba. This could only succeed provided the Turks had no idea what was being prepared. So the C-in-C, General Allenby, decided to mount a cavalry attack over exceedingly difficult country, an attack which they would not normally expect to be likely or possible. Two officers, Colonel Richard Meinertz-hagen of Allenby's intelligence staff and Colonel Archibald Wavell of the cavalry, were charged with deceiving the enemy.

The plan they devised was a classic. In case the enemy spotted signs of movement towards Beersheba, their idea was to convince him that these were connected with a feint attack, designed to distract Turkish attention from the real onslaught, which would be launched immediately afterwards at Gaza. But planting this idea in the enemy's mind had to be done very subtly: too obvious an attempt at deception would cause the Turks to realize what was happening. So the false information had to be planted as slowly and in as many different ways as possible, to make it appear as genuine as possible.

First of all Meinertzhagen left prepared evidence out in no-man's-land, where it would be picked up by an enemy patrol. This consisted of a staff officer's haversack, stained with blood (actually horse's blood) to look as if it had been dropped from weakness and wounds rather than as a deliberate plant. Its carefully chosen contents included details of the original target date for the real attack, in case the Turks had spies who might have found this out, but with more information which showed that the plan had now been changed to a *later* attack at Gaza, preceded by a feint attack near Beersheba intended to divert the Turkish defences. Other details were added to make the evidence more convincing—a letter from the officer's wife telling him of the birth of their child, some money and a book with details of a British cypher, all details which would never have been left behind willingly.

Meinertzhagen himself planted the evidence by taking the haversack into no-man's-land and deliberately bumping into a Turkish patrol, which opened fire on him. As soon as bullets came whining past him, he beat a careful retreat after leaving the bloodstained haversack, a water bottle and a pair of binoculars as if dropped in a desperate attempt at escape.

The obvious next step was for the Turks to listen to all British radio signals, in the hope of picking up messages using the captured cypher. So Meinertzhagen used this to feed doctored messages to the enemy—which included a reference to his own court martial for having lost the valuable haversack, and other pieces of circumstantial evidence which hinted at the authenticity of the material it contained, or which dropped, piece by piece, more information about the deception plan.

Finally, this gave way to psychological warfare of a more direct kind. Meinertzhagen was aware that the Turks were chronically short of tobacco, so he had more than a million cigarettes specially made and dropped over the enemy lines on the eve of the attack by British aircraft. The cigarettes were heavily doped with opium, and acted as a powerful soporific. In the meantime, the cavalry had moved inland from Gaza to Beersheba, its place being taken by dummy horses made of straw and its movement camouflaged by heavy fake wireless traffic in the Gaza area.

When the attack was launched, it achieved almost total surprise. The defenders of Beersheba were taken in their sleep by a hurricane bombardment, followed by an immediate cavalry onslaught. The Turkish lines crumbled, and the attack was followed by a second frontal blow at Gaza, which turned a collapse into a rout. Within six weeks, the British had taken Jerusalem and the campaign was all but over.

To return to the Western Front, when the breakthrough came in 1918, it was the Germans who first succeeded. They, too, had been busy developing new tactics to make the best use of the divisions brought in from the Eastern Front. These were well-drilled in mobile warfare and had evolved a style of fighting which was to be the basis of the blitzkrieg of a generation later.

The German offensive opened on 21 March against the over-extended British Fifth Army. The build-up for the attack had been carefully hidden to lull the British command into a false sense of security. Troop train movements gave no clue as to where and when the blow would fall, since the attacking units were only marched in by night at the last possible moment. The assault itself opened with a shattering barrage on the British model, using gas and smoke shells mixed with the high explosive to cause maximum confusion without breaking up the ground too much. The British in their turn had adopted the German ideas of defence in depth, with a Forward Zone, a Main Battle Zone and a Reserve Zone — but they had put too many men in the forward trenches, where they were overwhelmed by the shelling and the first waves of the attack.

For those British troops who survived the shells, the attack was an unnerving experience. Small parties of German shock troops pushed forward quickly through the smoke and the morning mists, dragging machine guns on sleds, infiltrating the British positions and choosing the lines of least resistance. Where defending troops held out, they found the Germans pushing past them and surrounding them, pinning them down until the artillery could finish them off. For men used to fighting shoulder to shoulder in trench lines, it was a difficult kind of fighting to counter. One by one the surviving positions were overrun or pounded into silence, and still the German advance went on. By the end of March it was halted forty miles further back, with a loss to the British of 178,000 men and more than a thousand guns – a crushing victory to the Germans after a long and bitter struggle.

Only later did the growing Allied superiority in numbers begin to tell. At Le Hamel on 4 July a brilliant set-piece battle showed what could be achieved when no efforts were spared to surprise and deceive the enemy. Allied aircraft had buzzed the front at low level to drown the noise of the tanks moving up for the attack. In just ninety-three minutes the Australians and Americans captured 1,472 Germans, two field guns and 171 machine-guns for the loss of 900 of their own men and three tanks damaged. Never had the value of surprise and planning been so tellingly demonstrated, and never had the troops' morale been so high during an attack. Even the wounded, carried forward on the backs of the tanks, had cheered the progress of the advance.

It was this winning formula which was to be applied on a much greater scale in the main British attack at Amiens. With eighteen divisions, more than 2,000 guns, 534 tanks and 800 aircraft, this was the most complex assault yet on the German lines. At long last the value of surprise and deception had reached the highest ranks of command, and no effort was spared to mislead and confuse the enemy before the attack began. Concealing all preparations of an attack anywhere along the front was virtually impossible. Instead, they opted for making the Germans think they had spotted attack preparations in another sector altogether. Once they were convinced the British would attack in Flanders, for example, they would be less inclined to watch for suspicious movements on the Somme front.

All the same, the greatest possible care was taken to drop the right information into the minds of the Germans. The first requirement was an attack which would be logical enough for the Germans to believe that what they saw was the truth. Foch, the Allied Commander-in-Chief, had been trying to persuade Haig to launch an attack against the German lines south of Ypres. Haig had turned it down in favour of an offensive against Amiens, fifty miles further south. But if the Ypres attack had seemed logical enough to the French, then the idea of an attack in this sector might well convince the Germans.

One obvious constituent of any attack was the fresh, well-equipped Canadian Corps; another was the cavalry which would, it was hoped, be ready for the breakout. Since the cavalry would be too far back to be obvious to the Germans, all that had to be done was to disguise the presence of the horse lines and supplies of forage in the Amiens area. But the Canadians were front-line troops, then serving in the First Army in the Arras sector. So the first step was to transfer two battalions of Canadians, together with a Canadian wireless section and two casualty clearing stations, north to the Ypres area, where they were put into the Second Army's lines opposite Kemmel Hill. The presence of the Canadians was made as obvious as possible, and the wireless section simulated a huge increase in coded wireless traffic in the area. At the same time a small squadron of tanks was sent to the Saint Pol railhead to make as much noise as possible behind the Ypres front, while the Royal Air Force stepped up the local air activity.

Meanwhile preparations for the real attack went ahead as stealthily as possible. The attack was due to open on 8 August, but no Canadian troops were moved into the sector until the day before, and they finally moved into their jumping-off positions just two hours before the offensive began. By that time all the other major preparations were finished. The civilian population had been evacuated from Amiens in April in any case, during the German offensives, and now the empty streets were covered with a thick layer of straw. Night after night the artillery trains rumbled through the town, the noise deadened by the straw-covered *pavé*—thirty brigades of heavy artillery and four siege batteries, guns, howitzers and ammunition limbers, all were in position and hidden from German eyes by first light each day. Huge dumps of ammunition were built up at night, to avoid the German Air Force; had the RAF swept the skies clear in this sector, that would have been another pointer to the real attack. Nothing was left unexplained, for the benefit of the Germans or any of their agents; when troops were sent forward ready for the attack, the official explanation was that they were moving to take over the defence of another sector of the front from the French.

On the night of the attack the final preparations began. The RAF concentrated in the area from all along the front, flying continuous night patrols to deaden the rumble of the tanks moving up to the front line, and the attacking troops moved into position with them. As the skies lightened, the whole strength of the RAF, now outnumbering the Germans seven to one, was hurled against the German airfields. With the sky clear of enemy fighters, British bombers could attack the German reserves moving up to meet the attack as they got off their trains in the stations at Peronne and Chaulnes, while the fighters could strafe the marching columns or the retreating defenders of the front lines.

Zero hour was twenty minutes past four on the morning of 8 August. An hour beforehand, the leading troops had crawled forward to within 100 yards of the German front line, waiting for the barrage to begin. When a thousand guns opened up through mist which cut visibility in some places to less than ten feet, the surprise was total. The dazed Germans found the barrage lifting to give away to lines of tanks and infantry bearing down on them. The speed and weight of the attack were too much for them and the lines bent and cracked, before giving way altogether in many places. By the middle of the morning, as the last mist dissolved in the warmer air, the Australians reached their final objectives and the Canadians were approaching theirs, after pushing for miles into the German defences. Victory was complete — more than 12,000 Germans captured and more than 300 guns, and an eleven-mile gap punched in the German lines.

Amiens did not produce a total German collapse; yet again they managed to stabilize the front and hold firm on a new defence line. But for all that it was a brilliant success by any standards, particularly those of the Western Front. The Germans may still have been unbeaten, but much of the old fight had been hammered out of their troops. The German Official History later referred to 8 August as the greatest defeat the German Army had suffered since the beginning of the war. Three days later, the morale of many of the retreating troops was so far gone that they cursed the reinforcements coming to join them as pawns who could only help to prolong the war. That same day Ludendorff offered the Kaiser his resignation. The Kaiser refused, but went on to admit that, 'We have nearly reached the limit of our powers of resistance. The war must be ended.'

By this time the effects of the iron grip of the British naval blockade were being felt with cruel intensity inside Germany itself. Food was desperately short; civilians were being called upon to make appalling sacrifices so that the army could be reasonably fed. Even then, conditions were bad enough compared with those which gave Allied troops cause to grumble. Morale in the German Army had taken an unexpected knock at the time of their greatest successes in the spring offensive of 1918, when captured Allied trenches had been found stocked with provisions which had been unobtainable luxuries in Germany for as long as anyone could remember. Now, as 1918 drew towards its close, the German population was fast approaching real starvation. Only the most rigid censorship prevented word of the catastrophic defeats now being suffered by the Army for whom so much had been sacrificed from reaching the debilitated and dispirited population at home.

The newly-constituted Allied propaganda teams were now to strike with deadly effect. The idea of a propaganda offensive against Germany had been suggested as far back as 1915, but had been turned down at the time on the grounds that mounting an attack in propaganda terms would be taken as an admission of weakness. But in 1918 the Minister of Information, Lord Beaverbrook, appointed his fellow newspaper magnate, Lord Northcliffe, to be Director of Propaganda in Enemy Countries. From their headquarters at Crewe House, Northcliffe's men issued a stream of information aimed at German fears and susceptibilities. Aided by writers of the calibre of H. G. Wells and Rudyard Kipling, the British propaganda effort blazed a totally new trail, lighting the way for much of the Allied propaganda offensive in the Second World War.

Methods of delivery were still fairly primitive. To reach the German troops in the trenches, propaganda leaflets were delivered by shells, balloons and aeroplanes. Most of this material was so-called 'white' propaganda – in other words, there was no real attempt to disguise the fact that it had been produced by the Allies. But it obeyed two of the classic rules of good propaganda – it was sympathetic in tone to the predicament of the Germans, soldiers and civilians alike, and it stuck closely to the truth. For with the situation as it was in 1918, there was no need to invent or distort what was happening; the truth was devastating enough. The leaflets showed accurately the latest Allied gains, and quoted German casualty figures. The experiences of the front-line soldiers showed them that what the leaflets said was accurate enough. At the same time, the rest of the theme – that it was high time the German people took control of events into their own hands to arrange an armistice as speedily as possible and end the suffering imposed on them – was given extra weight by the accuracy of the factual parts of the message.

Meanwhile long-range aircraft were dropping leaflets over Germany itself, up to 167,000 a day, carrying the same message – that Germany's leaders were prolonging the senseless slaughter for their own ends, and that the longer Germany continued the struggle, the greater the suffering would be, and the harsher the peace terms which would eventually be imposed. But there was a taste of the carrot as well. Each leaflet reminded its readers that a truly democratic Germany would find a place in Europe after the fighting was over, provided something was done quickly.

Some propaganda efforts were more cruel. A list was circulated around Germany's ports entitled 'The Lost U-boat Commanders'. Using detailed information from Allied naval records, it told of the fate of 150 German U-boats, their crews and commanding officers, who had failed to return from war patrols. Other items laid the foundations of what later came to be called 'black' propaganda, propaganda which disguised where it came from. Many Germans, to avoid their own rigid censorship, went to great lengths to read papers from neutral countries, which still had access to at least some of the war news. The Allied propaganda teams had no control over what these papers published, but a barrage of readers' letters, written in Crewe House, but posted through embassies in neutral capitals, was aimed at the offices of these papers. The tone of the letters was thoroughly pro-German, to avoid their being stopped by the censors, but somewhere in each would be an item contrasting the conditions under which the brave Germans were fighting, compared with the luxury of life in Britain or France. Drop by drop, the truth of the message filtered through into the minds of German readers. Being passed onwards by word of mouth, the message was far more effective.

Books and papers, pamphlets and leaflets were smuggled into Germany, and among the troops in the trenches appeared new and much more outspoken editions of their own trench newspapers. These too were Allied creations, carefully written and matched in layout and typeface to the originals. More than half a million a week were printed at the height of the propaganda offensive and dropped just behind the German lines by Allied aircraft.

The effect of this Allied-aided dawning of the full truth on the German people was literally overwhelming. Civil unrest was mounting and the army itself was weary of the war. Yet demoralized as it was, the German army was still a formidable enemy, and Allied opinions were sharply divided as to how to achieve total victory. The British and French commanders blanched at the thought of another year of fighting. So, except for Pershing, the American Commander-in-Chief, who pressed for nothing less than total and unconditional German capitulation, counsels only differed on the terms needed for an armistice. Too lenient terms would be a betrayal of all the years of fighting and the millions who had died, and the spectre of another attack by a resurgent German Army after time for rest and regrouping would always be present. Yet terms which were too harsh would keep the Germans fighting to the bitter end. What was to be done?

Fortunately, by late 1918, the truth of the Allied propaganda was beginning to produce real changes. Germany at that time was an ideal target for psychological warfare; once people accepted the facts the Allies stated, and which were all too often borne out in detail by soldiers returning from the front as wounded or, increasingly, as deserters, then the rest of the message tended to be absorbed as well. The German leaders were tainted with the accusation that they were keeping the agony going for their own ends. So the interests of the Allies and the German people coincided exactly. Neither wanted another year of war; both knew it would mean another heavy toll which could and should be avoided.

So the façade of Imperial Germany began at last to crack. The navy, face with orders to sail against what they now firmly believed to be an invincible enemy, mutinied. Ludendorff resigned, this time successfully, and the newly-appointed Chancellor, Prince Max of Baden, sent a proposal for an armistice to the American President, Woodrow Wilson. But while the overtures for peace were being played, the war machine threatened to blot them out by continuing exactly as before. Although Germany's allies were surrendering one by one, German U-boats were still sinking ships without warning, and German troops in France and Belgium were still destroying property and carrying away hostages as they retreated.

But the people at home were beginning to take control. Dock and factory workers in the ports sided with the naval mutineers, and troops in the home garrisons, reinforced by increasing floods of deserters from the front, were spreading the unrest. Desperate measures were needed, and, on 9 November, Prince Max tried to pre-empt the revolutionaries by announcing, without authority, that the Kaiser had abdicated. He proposed a Socialist politician, Friedrich Ebert, to succeed him as first Chancellor of the new Republic, and called for a general election.

Two days later it was all over. The guns fell silent and the weary troops on both sides tried to accustom themselves to a life expectancy of years rather than days. But the fact that the civilians had taken the initiative in pushing for an armistice was to lay the foundations for another catastrophic war a generation later. For although November, 1918, had seen the decisive defeat of the German Army on the field of battle, the way in which peace had been declared tended to blur this clear and simple fact. After four years and three months of war, the Army was still in possession of French and Belgian territory, and it was still—just—a united and disciplined force. When at last it returned home, it did so with colours flying and bands playing—scarcely the appearance of a defeated army.

Within months of the signing of the armistice, the legend was spreading that the Army had been stabbed in the back by plotting civilians and cowardly politicans. In time it was to produce a national longing for a resurgent army and a deep distrust of democratic government, which paved the way for an even more catastrophic war only two decades later.

3

In the early days of the war Field-Marshal Hindenburg had said that the side with the strongest nerves would win. The Allies had proved him right, but the circumstances of their victory meant that already they were beginning to lose the peace which followed.

CHAPTER TWO

BUILD-UP TO A NEW WAR

FOR TEN years and more war-weary Europe maintained an uneasy peace. But behind the frivolous façade of the nineteen-twenties, dark forces were already astir in the soul of Germany. The end of the war had left deep and bitter wounds. The post-war republican governments had never had true popular support. The system of proportional representation produced a great number of evenly matched parties, none of whom were strong enough to command an overall majority in the Reichstag. Weak coalitions slid from crisis to crisis, dissolving and reforming; the default in the heavy reparations payments due to the Allies, the resulting French seizure of the Ruhr, the strikes and the collapse of the mark, together with the ever-upward climb in the numbers of the unemployed, all conspired to build up a tidal wave of bafflement, resentment and fear, a wave which was to sweep into power a failed Austrian artist, an ex-corporal in the Bavarian infantry, leader of a party of small-time politicians and rough-necks, dignified by the title of the German National Socialist Workers' Party, an indigestible mouthful better known by its German contraction—the Nazis.

Adolf Hitler seemed to many people outside Germany to be a ludicrous figure. But he possessed a brilliant insight into the minds of his own countrymen and into those of the leaders of other countries. He had a knack of exploiting Germany's very real distress to aid his own climb to power and once he reached his initial objective, he embarked on a career of conquest backed by a brilliant exploitation of psychological pressure.

First he needed the teeth to give his threats the background of force they needed. Clauses in the Versailles Treaty allowed Germany an army of up to 100,000 men but no general staff and no tanks, a small navy with no ships larger than 10,000 tons displacement, no submarines, no air force at all. One day, Hitler knew, he would be strong enough to declare the treaty restrictions null and void, but in the meantime he was in a chicken-and-egg situation. He had to break the treaty in secret in order to build up the forces which would allow him to tear it up in public. So Nazi Germany embarked on the first of the great deception campaigns of the Second World War—one which gave birth to the mightiest war machine the world had yet seen and made war not only likely but inevitable.

Expanding the army was perhaps the easiest task, since the nucleus already existed. Every precaution was taken; orders were given verbally, figures were never put in writing, and publication of the normal annual Officers' List was cancelled lest the increased numbers gave the game away. In just eighteen months, the target was a threefold expansion to 300,000 men, and this was merely the beginning.

Krupps had been hard at work on guns and tank design since the middle 'twenties and the components for the first half-dozen of Germany's foreign-built U-boats were stored at Keil, waiting to be assembled. At the same time Germany's shipbuilders were perfecting their own techniques by building submarines for foreign customers, such as Finland and Turkey. Before these were handed over to their new owners, each one was tested long and thoroughly by officers and men from Germany's own 'Anti-Submarine School'—a cover for what was in reality the nucleus of the new U-boat arm undergoing its first vital training in submarine operations and tactics.

From this point onwards, Germany's strength grew by leaps and bounds. In 1936 Hitler resolved to bury another legacy of Versailles—the demilitarized Rhineland. Without warning, on 2 March, 1936, he sent a series of small detachments of German troops across the Rhine bridges to take possession of Germany's westernmost province. With bayonets fixed, bands playing and standards flying, they made a brave sight. The flabbergasted French later estimated that three German divisions, or 40,000 men were involved. In fact Hitler had sent only three battalions across the river. Had any one of the thirteen French divisions waiting on the other side of the frontier moved against them, the German forces would have been overwhelmed. But such was Hitler's reputation for toughness and the increasing prestige of the German armed forces that no one moved to stop him. From now on he would repeat the process to win whatever territory he wanted—and with every month that passed, the increasing power and threat presented by the German forces would make the task of retaliation both more dangerous and less likely.

By 1938, Hitler was ready to absorb his old homeland, Austria. On 12 March, 1938, after months of increasingly virulent threats, he sent the *Wehrmacht* across the border. There was no opposition; his troops entered Vienna in triumph. Yet already the first ominous cracks were appearing in the invincible war machine. The panzer forces were showing disturbing signs of wear and tear, after what had been a simple, unopposed drive of a few hundred miles along well-made roads. So badly organized was the supply system that the men had to fill up their vehicles by seizing stocks from roadside service stations. By the time they reached Vienna at the end of the day, between a third and three-quarters of the German tanks (depending on whether one chooses to believe General Guderian or General Jodl) lay stranded by the roadside in the wake of the advance.

At the same time, in order to raise enough men to take over a totally recumbent Austria, Hitler had been compelled to strip the frontier with France of most of its defenders, at a time when France had one of the largest standing armies in Europe in a high state of readiness.

Goebbels' propaganda machine had dwelt at great length on the impregnable fortifications of the Siegfried Line, which guarded Germany's western frontier, but in fact the Line was a sham. However, because the French had expended so much energy on their own much stronger Maginot Line defences, they were all too ready to believe that the Germans were capable of something even more formidable. The trench-system deadlock of the First World War still exerted a powerful effect on military thinking, and the idea of uncrushable German defences was a heaven-sent alibi for taking no risks. The outside world saw only the newsreel films of the small sectors of the Siegfried fortifications which *had* actually been built, just as it was shown the lines of tanks which *did* succeed in finishing the drive to Vienna. So the reputation of Hitler's forces climbed another notch in the estimation of the world.

From now on, however, he was to find aggression more difficult to justify. The Rhineland had at least been part of Germany, and Austria was a wholly German nation. Next on his list was Czechoslovakia, a country defended all along its border with Germany by a line far stronger than the mythical Siegfried defences. There were massive fortifications buried deep in the Bohemian mountains, backed by an army which had a million men on call and which, together with those of Czechoslovakia's ally, France, outnumbered the total German forces by more than two to one.

Yet once again Hitler had his way. He seized on the Sudeten Germans, a minority remnant from the old Austrian Empire, who still lived in the border areas facing German territory. By a carefully orchestrated scheme of propaganda, these Czech Germans were induced to demand that their part of Czechoslovakia be handed over to Germany. Hitler himself backed their demands by more threats of his own. So powerful was the psychological combination of his protestations for peace and the fear of his apparently limitless armed strength that his vastly more powerful neighbours handed over the Sudeten Germans and their borderlands to Germany.

With them, as Hitler had intended, went the vital Czech border defences, which he would never have taken by military action. After the crisis meetings in Munich in September, 1938, he was given all the Czechs' trump cards without a shot being fired. Six months later he took over all the rest of Bohemia and Moravia by simply marching in and occupying it. One bonus for the German dictator was the capture of almost twice as many battleworthy tanks as his own factories had so far been able to deliver.

Poland, next course on Hitler's territorial banquet, promised to be more difficult to swallow. Alarmed by the fate of Czechoslovakia, Britain and France threatened to go to war if Poland were attacked. Hitler, for his part, reckoned that all that was needed was a little more care over his psychological preparations, and once again he would gain all his objectives.

First, Goebbels' Propaganda Ministry stepped up a campaign against Poland, dwelling on the alleged persecution of Germans in Polish territory. Groups of 'refugees' were assembled to tell the world's press of their flight from Polish cruelty, and more and more atrocity stories were circulated. But the keystone of the deception programme was an operation bearing the title 'Canned Goods', intended to provide a solid and completely irrefutable example of blatant Polish aggression.

The plan was to be carried out under the command of a young SS officer called Alfred Naujocks, who was ordered to simulate a Polish attack on the German radio transmitter at Gleiwitz, close to the Polish border. His men were dressed in Polish uniforms, and when the attack took place, on 31 August, 1939, they 'seized' the radio station and broadcast a message in Polish full of apparent defiance and hatred for the Germans, proclaiming that Poland was now declaring war against Germany. They were then 'driven off' by loyal German troops, but one extra macabre detail was left on the scene as evidence. Concentration-camp prisoners had been dressed in Polish uniforms, heavily sedated and then shot. Their bodies were scattered around the radio station to convince otherwise sceptical pressmen from neutral newspapers that the attack had indeed been genuine.

At that time the true explanation was harder to swallow. Not everyone

realized that the Nazis were perfectly capable of murdering helpless
prisoners in cold blood as part of a propaganda exercise, and the result
was that many observers believed the unmistakable evidence of their
eyes, a powerful demonstration of the effectiveness of a totally unscru-
pulous lie. Hitler now believed he had the justification he needed, and
at dawn the following morning the *Wehrmacht* crossed the Polish
frontier.

For once he was to be surprised by the reaction. Within two days
Britain and France had declared war. Yet the shrewd psychology and
the cunning deceptions which had brought him so many triumphs
during the long build-up to war continued to bring him his greatest
military victories. Even as Poland was being swiftly and efficiently
dismembered by the German armed forces, the Allies remained
paralysed by the myth of German invincibility, while the Reich lay
wide open to any determined attack from the west.

This was a direct consequence of the Germans' pre-war psychological
coups. So well entrenched was the idea of the invincibility of the
Wehrmacht that it totally distorted Anglo-French intelligence estimates.
In the first few weeks of the war, with the *Wehrmacht* fully committed
in Poland, the Poles themselves told the Western Allies that on the
basis of the divisions they had identified attacking them, there could
only be at best twenty-five to twenty-eight German divisions on the
whole western front. British Intelligence thought this an over-optimistic
estimate, and plumped for between thirty and thirty-five. The French,
although they had seventy-two divisions formed and in position the
week before the war began, thought the situation was even more
unfavourable. They rated the total German strength at 135 divisions,
fifty per cent higher than it actually was, and French Intelligence
claimed to have identified twenty-six out of a possible forty-three
German divisions manning the Siegfried Line defences alone.

It was all an illusion. Hitler was still convinced that the Western
Allies would not risk an attack to save the Poles and he planned his
troop movements on that basis. Tragically, he was right. He was able,
for example, to guard the whole Belgo-Dutch border with just one
operational division, backed up by two more raw and untrained divisions
and a pair of Home Guard divisions. The French commander on the
Western Front had six times as many guns as his German opposite
number; and, with all the Panzer divisions on the eastern front, the
Germans had no tanks, while the French had more than three thousand
ready for action.

So for eight months German psychological propaganda won their

army a respite, so that it could rest and refit after the exertions of the
Polish victory, regroup and move westwards. That delay gained the
Allied armies nothing, for, as early as October, 1939, Hitler was drawing
up plans for their destruction. At first this was to be a classic German
attack across Belgium, a remake of the 1914 epic, but this time with
the shatteringly effective combination of tanks and dive-bombers,
backed up by the onslaught of a highly trained mechanized army.

This was more or less what the Allies expected and the kind of attack
their plans were intended to deal with. As the Germans advanced into
Belgium, they were to be met with a similar Allied advance to stop them
along the same river-line defences which had seen the battles around
Mons in 1914. The Allied armies had the advantage of numbers; even in
the decisive part of the front from the end of the Maginot Line northwards, they mustered fifty-nine divisions against fifty-five German when
the invasion *did* come. The much-vaunted German tank superiority was
another myth. In the main the Panzer divisions were forced to rely on
the outdated Mark I and Mark II light tanks. Between them they could
only muster 627 Mark III and Mark IV medium tanks, and 381 Czech
38t tanks which they had taken over after the annexation of Czechoslovakia. The French Army could muster 3,000 tanks, admittedly some of
dubious quality, but generally speaking the Allied tanks were as well
armed and better armoured than were the Germans.

But the Germans had a priceless extra ingredient to add to their bold
planning – they had the experience of using tanks in action. And, apart
from the psychological reward of practical experience, the Polish war
had taught them valuable lessons. Before that invasion the Germans
had tended to believe, along with everyone else, that only flat, open
country was suitable for tank warfare. In fact their Panzer formations
had operated in Poland with great success through country they had
previously thought impassable to tanks.

German tactical thinking favoured grouping the bulk of their armoured forces to deliver one shattering blow at a single particular sector
of the enemy line, then pouring through the gap as quickly as possible
and fanning out into the open country beyond to exploit the breakthrough. But where should the blow fall? The Allies had the dilemma of
all defending forces – *their* tanks would have to be spread right across
the likely areas of attack. The Germans could pick whatever sector
suited their plans best.

The idea of using the entire German tank force to deliver one
paralysing blow came from General von Manstein, Chief-of-Staff to
General von Rundstedt, commander of Army Group A. Drawing on

his Polish experience, von Manstein was convinced that this would achieve the greatest psychological success and the biggest surprise if it were delivered on the one sector of the front where the French did not expect a tank attack. In other words, the Germans should invade through the Ardennes.

Events soon played into von Manstein's hands. The original High Command plan for a frontal assault through Belgium was in the possession of an officer flying to one of the army headquarters when his plane made a forced landing on Belgian soil. The officer was arrested before he could destroy the papers, so the German General Staff decided that that plan would now become a deception. The Allies would be left to believe that the Germans were making their attack according to the original schedule – but, once the Allied forces were spread out across Belgium, the main blow would be launched, according to Manstein's prescription, through the Ardennes.

And so it transpired. While all eyes were on the northern end of the front, the tanks for the main assault were assembling in the gloomy dampness of the Eifel forests, ready for their advance into the Ardennes. Every precaution had been taken to avoid alerting Allied Intelligence, such as cutting signals traffic to the absolute minimum, so that the attack eventually came as an almost total surprise.

The German armour carved its way through the Allied line with little difficulty, yet, even so, the attack demanded total psychological mastery. As the leading Panzer divisions thundered towards the Channel, they unrolled an exposed flank which was highly vulnerable to counter-attack. Really determined resistance could still have upset the pace and balance of the German advance, even if halting it altogether was now too much to hope for. And when the Matilda tanks of the British 1st Armoured Division mounted a determined if limited attack against the flanks of Rommel's Seventh Panzer divisions west of Arras on 21 May, the result was a severe shock to the Germans.

As usual the armoured regiment at the head of the divisional column was far ahead of the rest of the troops, and the British tanks ran into the infantry who were following up. These were protected by German 37 mm anti-tank guns, but to their horror the Germans found that their shells made no impression whatever on the British armour. The Matildas tore into the infantry columns, causing heavy losses and a panic-stricken retreat. Only the fortuitous presence of a troop of 88 mm anti-aircraft guns, which were turned into impromptu anti-tank guns, prevented a total rout. Realizing the danger, Rommel swung his Panzer regiments round to retrace their line of advance and come to the rescue of his stricken infantry, but they ran headlong on to a screen of British anti-tank guns, who knocked out twenty armoured vehicles.

The effect of this sharp little battle was negligible in terms of stopping the German advance, but its psychological effects were tremendous. To the British it showed that German tanks, and German tactics, could be beaten by sound thinking and resolute action. To the Germans it was an unwelcome reminder of how easily the tables could be turned. For Rommel it was a lesson he never forgot in how to employ tanks and anti-tank guns in concert; for the rest of the Panzer commanders it was a reminder not to accept battle on the enemy's terms.

This may well have been the reason for the halt in the headlong German advance as the British and the northern French armies fell back on Dunkirk and the Channel. Defence positions were as strong as could be prepared in the time available, with dug-in gun positions, flooded fields and anti-tank ditches—far from ideal Panzer country. And the tanks themselves needed time to refit and repair after their arduous campaign. But perhaps the biggest factor in the German hesitancy to finish off their enemies was their attitude to the Channel. To a Continental people, to an army facing the probability of having to cross the water to mount a full-scale seaborne landing, it was a daunting obstacle. To the naval-minded British the water was the route to home and safety, and no orthodox general should ever be blamed for not foreseeing the armada of little ships which helped to make Dunkirk possible.

By the middle of June, 1940, the British Expeditionary Force, or most of it, was back on British soil, and the French were well on the way to total defeat. All the same, the British position was desperate. All the equipment of the BEF had been abandoned in France and little existed to stop the Germans should they launch an invasion. Civilian morale could be kept up by speeches, news broadcasts, and the raising of the volunteer Home Guard to watch for parachutists and Fifth Column- ist activities. Apart from this, all that could be done from a psychological point of view was to try by every possible means to deter the Germans from crossing the Channel, at the very least until some sort of defence could be organized.

Saying that the British were ready to repel an invasion was one thing— the Germans would expect British propaganda to say that, true or not. In fact, if the British really were that confident of beating off an attack, then a stronger argument might be that they would keep quiet, in the hope of luring Hitler's forces into a bloody repulse. On the other hand, the Germans would hardly doubt the evidence of their own eyes;

troops, tanks, artillery and supply dumps were hard to hide from reconnaissance aircraft, however carefully camouflaged. Trained observers could quickly pick out the tell-tale signs of military occupation. So the Germans were bound to mount a full-scale reconnaissance of the potential invasion coast; and when they did, it was vital that there should be something for them to see.

Training men and building equipment, even on a modest scale, would take far too long. Instead, the invasion defences had to be made from whatever was to hand, as convincing yet as insubstantial as theatrical scenery. Men were represented by cloth and cardboard dummies; tanks, guns and lorries by folding assemblies of cloth, wood and cardboard; dummy tank-traps were made from wood, dummy pill-boxes were put up alongside imitation slit-trenches, and the whole area covered with vehicle tracks. From the cockpit of a speeding bomber, several hundred feet up at two hundred miles an hour, as likely as not being harrassed by defending fighters, the illusion was real enough, and would be confirmed by photographs.

When the Germans returned to the area to check again, the tanks and men would be gone, taken to pieces, loaded on lorries and sent to another part of the coast to do their duty all over again. In their place would be a scene as tranquil as it had been before their arrival. But, to the trained eyes of the German photo-interpreters, there would be subtle signs of badly blended camouflage, hinting that all the men and equipment previously in open view were still there in hiding. German military intelligence departments recorded the steady build-up of strength all along the southern and eastern coasts of England. As time went by, the camouflaged models themselves became more sophisticated — 'men' were mass-produced, moulded in rubber or plaster, and some improved models were wire-operated to add a touch of movement to an otherwise static tableau.

There were other ploys in the battle to convince the Germans that any attempt at invasion would meet a strong and effective response. Experiments had been carried out to try laying a carpet of fire on the surface of the Channel by pumping a thick coating of oil on to the water and then setting light to it electrically. In fact this would have been crippling to our meagre oil reserves. But the threat of fire on such a scale was a splendid psychological weapon, and the Germans could hardly avoid seeing signs of the preparations from across the Channel. From the French coast the towering walls of flame produced by the tests could clearly be seen and German patrol boats reported similar tests off East Anglia. Rumours of this 'secret weapon' were fed to neutral papers and dropped into British propaganda broadcasts to the Germans, For example, Sefton Delmer ran a series of English lessons for German listeners on the BBC (see Chapter 5), one of which was angled specifically to the *Herren Engelandfahrer* about to make the *Kanalüberfahrt* (Channel crossing). He highlighted useful phrases like *Das Boot sinkt*, (the boat is sinking), or *Das Wasser ist kalt*, (the water is cold). Delmer went on to instruct his listeners how to conjugate the verb to burn: 'I burn, you burn, he burns', and so on, ending with some useful everyday phrases using these words, such as *Der SS Sturmführer brennt auch ganz schön* (The SS Captain is also burning quite nicely).

Undoubtedly the Germans themselves were worried about the possibility of burning oil engulfing their landing barges. So worried, in fact, that they tried proofing the barges with sheets of asbestos. They tested the idea on some experimental barges loaded with troops, which were then sailed through a blaze intended to simulate the British defences. Unfortunately they had under-estimated the power of the fire and most of the troops were burned to death and the barges destroyed. For days afterwards charred corpses were said to have been washed up on both sides of the Channel. The Political Warfare Executive went into action, spreading the story among neutral and occupied peoples that the true origin of these bodies was that the Germans had indeed attempted a landing, but had been beaten off with heavy losses.

* * *

As the invasion threat faded, through the autumn of 1940, land fighting broke out against a different enemy in a different theatre of war. The Italian Army in North Africa began preparations to invade Egypt, from whence small British forces had been mounting a series of sharp nuisance raids against the Italian colony of Tripolitania. The attack began on 13 September, and a whole Italian army moved slowly and ponderously forward, taking four days to travel sixty miles against negligible opposition. It stopped at a settlement called Sidi Barrani. A series of camps were built around the perimeter of the Sidi Barrani position; and there the army settled down to wait and find out what kind of reply the much weaker British forces had in mind.

The desert was a totally new theatre of war, imposing its own demands and its own challenges. Apart from the awesome task of having to ferry every can of fuel, every scrap of food and every drop of water across distances of hundreds of miles of scorching sand, any normal concept of lines of defence or of occupying ground was totally out of place. The inland flank of any position in the desert rested on open sand, and instead of formal set-piece battles, the armies blundered into one another, more often than not, like a pair of blindfolded boxers in a darkened room. Units moved like ships at sea, alert for signs of the enemy at any point around their horizon, and depended for their existence on making regular rendezvous with their own supply columns.

This new kind of fighting demanded a whole new set of rules. Concealment was both total and non-existent. A whole army could be just beyond the horizon in any direction, without the slightest hint of its presence. But, within the circle of one's own vision, a single man or one lone vehicle stood out like a fly on a white wall. At the distances over which the fighting took place there was no way in which a tank or a gun could be hidden on the flat, open desert, and surprise became a rarely enjoyed tactical luxury.

Fortunately General Wavell, who commanded the British forces in Egypt, was a firm believer in the values of deception and surprise in winning battles against numerically superior opposition. Before the war he had written a paper on the principles of successful deception — including the need for total security and total plausibility, and the use of multiple bluffs — which was to lay the foundations for later Allied successes. He set up a special unit at his Cairo headquarters, under the command of his old friend and former colleague, Colonel Dudley Clarke, with a staff including the brilliant stage magician, Jasper Maskelyne. As Wavell well knew, what couldn't be hidden could be disguised, and confusion and illusion could be used as surely as shot and shell to demoralize and defeat the enemy.

The desert deception unit began in a small way. The initial Italian threat to Egypt had been from the air, since the balance of power was tipped heavily in favour of the *Regia Aeronautica*. Maskelyne carried out a series of experiments, using batteries of carefully-angled mirrors to concentrate search-light beams on the pilots of bomber aircraft so as to distort their sense of direction and perspective. Trials using volunteer RAF pilots had some spectacular results. The totally disoriented pilots had great difficulty in keeping flying at all, let alone in delivering attacks.

But perhaps Maskelyne's greatest air-defence coup was in the camouflage of the oil storage tanks in the Egyptian base areas. These were prime targets for German and Italian bombers. The Germans were by now experimenting with infra-red film which showed up camouflage quite clearly, so Maskelyne determined to throw the bomb-aimers off by other methods. Since the raids were invariably made by night, he created a distorting effect with searchlight beams which altered the perspective of the tanks themselves. This caused the bomb-aimers to release their bombs at the wrong moment, so that all the attacks missed. When reconnaissance planes came over to check the damage, enough faked destruction covered the area to make the Axis air forces consider the sites not worth a return visit.

Wavell's immediate problem was to persuade the Italians to keep their distance. Starved of tanks, the decisive weapon in desert fighting, how could the British keep their enemies from a head-on attack which could have only one ending? The deception unit was put to work. Lorries were made to look like tanks with covers of canvas, cardboard and sailcloth. But by this time it had become apparent that the Italians were far too cautious to try a direct attack, so the new problem was to disguise what little striking power Wavell had, and thereby lull the enemy into a sense of false security. This meant that the opposite approach was needed. Tanks must appear as innocuous and defenceless as lorries. So the deception unit came up with a fake lorry cab and tarpaulin cover which could fit over the tank chassis and a length of spiky chain-mail which could be towed behind the tank to stop the tracks left on the sand from giving the game away. At the same time, the rest of the army was defended by fake tanks, provided with convincing gun-flashes. Four empty oil drums, a pair of poles and a sheet of camouflage netting provided a convincing gun emplacement, backed up by the right kind of pyrotechnics. The Italians hardly expected the unpredictable British to take the trouble to import scarce materials merely to cover up patches of empty sand. It was far easier to take these camouflaged gunpits at face value. Even if the idea of deception entered their heads, innate caution reinforced the instinct to give them a wide berth.

By now the Italians were jittery and inclined to flight. A squadron of Valentines or Matildas had only to appear over the horizon for the enemy to take to their heels — satisfying for British morale, but no way to win battles. Only by taking on the role of a humble and vulnerable supply column could British armour be sure of luring the Italians close enough to hammer home a solid blow. Tanks were still desperately scarce and such tactics cut down fruitless chases and unnecessary wear and tear. At the same time the three-ton trucks could cross the desert almost with impunity. Canvas screens and wooden props changed their silhouettes to the menacing bulk of tanks, while the dust plumes thrown up by every vehicle crossing the desert disguised the presence of wheels pretending to be caterpillar tracks. Other tricks included inflatable cruiser tanks, false roads and mounted Arabs towing harrows to raise convincing clouds of dust which screened the detail of these deceptions. AA fire kept enemy aircraft high enough to be deceived by what they were apparently seeing.

Tactics as flexible as these proved invaluable in the advances following the victory at Sidi Barrani. Time and again the Italians were made to retreat by columns of disguised lorries, and there were still cases of enemy tanks being lured into battle by disguised Matildas, although the enemy was learning to treat *all* enemy vehicles with deep suspicion.

But this was just the beginning. How much more could be achieved by deception was demonstrated in a splendid set-piece mounted early in 1941. The first of the private armies of the desert, the Long Range Desert Group, had been formed on Wavell's orders by a physicist and long-time desert explorer called Bagnold. His brief was to make up for the desert army's chronic lack of reconnaissance aircraft by keeping watch over the vulnerable southern flank. As Bagnold's men became more adept at crossing the sterile wastes of the Great Sand Sea, navigating by stars at night and sun-compass by day, the possibilities began to swim into focus. They shot up the Italian staging post for their communications between North Africa and Abyssinia. They ambushed Italian convoys, snooped on traffic movements, intercepted messages and began to probe deeper and deeper behind the Italian lines.

But there was one major obstacle—the Italian garrison at Siwa oasis, two hundred miles south of the coastal road on the northern rim of the Great Sand Sea, and the crossing point for many of the desert trails and caravan routes. Avoiding it by a margin large enough to be safe added hundreds of miles to every journey. The capture of Siwa became a priority.

Fortunately the Long Range Desert Group, of all the units of the desert army, was the readiest to appreciate the value of unorthodox tactics. So when Wavell's deception unit, headed by Colonel Clarke, offered to help, the offer was eagerly accepted. At last light the assault on Siwa began.

As the light faded a flight of RAF Hudsons opened the attack, roaring over the oasis and showering the whole area with brilliant flares. Mixed with the flares were specially doctored fireworks which burst on landing and fired volleys of coloured lights into the air. A second wave of aircraft then dropped strings of dummy paratroops, weighted straw figures in full harness, also loaded with fireworks which went off in quick succession, like bursts of machine-gun fire. The effect on the Italians was shattering. The lurid glare of the flares, the splashes of the signal lights and the hundreds of figures drifting down under parachutes, spitting volleys all around them, convinced them that not only were they under a full-scale attack, but also that they were heavily outnumbered. Without staying to investigate further, they scrambled aboard their vehicles and roared off into the night. The LRDG had not had to risk a single man in the attack; nor did they lose a night's sleep or expend a round of real ammunition. They simply marched into the oasis next morning to find the Italians gone.

The Germans, when they arrived in the desert to assist their Italian allies, proved to be a tougher proposition altogether. To begin with, they themselves knew a great deal about the value of deception. When Rommel paraded the first units of the Afrika Korps at a celebratory march-past in Tripoli, he decided to impress the Italians, the local population and the Allied agents he knew must be watching. So squadron after squadron of German tanks squealed and clattered past the generals on the saluting base, then swung away down side-streets. The Panzer crews doubled back, re-appeared on the main boulevard and passed the saluting base a second, third or fourth time.

Deceiving the British on the field of battle was to prove much harder. But Rommel was fortunate enough to have the services of some deception experts of his own. Always short of tanks, the Afrika Korps turned lorries into apparent armoured vehicles and even *Kubelwagen* jeeps into armoured cars. Because they knew that many of their own vehicles were so disguised, the Germans were reluctant to take anything at face value. No longer could trucks masquerading as tanks put the enemy to flight by their mere appearance. The German reaction was to shoot at tanks or lorries alike, and any deception was all too quickly exposed.

This led to a new type of deception. Real tanks were mixed with disguised lorries so that no attacker could be sure of the strength of the force he was about to encounter. And with both armies relying on all kinds of captured enemy vehicles, repaired and put back into commission, and with the vastness of the desert always ready to provide a cloak of anonymity and uncertainty, North Africa proved an ideal testing ground for ideas of this kind. But Rommel had other aids, among them Hauptmann Alfred Seeböhm and the *Fernmeldeaufklarungs-kompanie*. This unit picked up almost all the British forces' radio traffic, and by very careful analysis was able to tie call-signs to particular units, to calculate where concentrations of troops and tanks were taking place by careful direction-finding, and to crack sufficient of the British codes to give Rommel a constantly up-to-date picture of

British movements and intentions. This enabled him to anticipate moves against him, and frequently to catch the British off balance at their weakest point. Shrewd and brilliant as Rommel's tactical senses were, Seeböhm's interception and analysis of radio traffic went a long way to ensure his enduring reputation

However, some British deceptions still worked, mainly those not tied to radio signals. Desert airstrips were vital and vulnerable, and almost impossible to conceal. But enough fake airstrips could be produced to make the task of spotting the real ones increasingly difficult. This phase of the war reached its ultimate development when the Italians actually raided a dummy airfield (which they had identified as such) with dummy bombs carrying rude slogans in English! But a far more effective measure was to cover the real airstrips with fake bomb damage, making the enemy think they were still unusable and not worth another raid. During the long siege of Tobruk this was applied to other vital places, like the Italian-built distillery which supplied the whole garrison with fresh water. The enemy bombers attacked this crucial target one night, but no real damage was done. Immediately the deception teams set to work digging fake bomb craters, the sides of which were darkened with oil and coal dust to make them look deep. An unused cooling tower was blown up to look as if bombs had destroyed it, debris was scattered everywhere and the main building had its roof painted to look as if a hole had been blown in the side from bombs exploding within. The next day German photo-reconnaissance planes came over the distillery, after which it was left alone throughout the siege.

Yet deception really came into its own, paradoxically, when the desert proper was left behind. In the summer of 1942 Rommel gained his greatest success with the capture of Tobruk, largely thanks to Seeböhm's interception of signals referring to British counter-attacks. Rommel attacked just before the British attack was due to begin, caught the Allied forces off balance, and took the fortress. But as he pushed the British back towards the Nile disaster struck. British direction-finders tracked down the position of Captain Seeböhm's unit to a site identified as Hill 33, at the northern end of the German front at Alamein. A stealthy raid by a battalion of Australian infantry achieved maximum surprise: an Italian unit was captured in its sleep, and the first warning Seeböhm's men had was the landing of smoke shells from the dawn barrage, followed by the arrival of the Australians with fixed bayonets from out of the encircling gloom. Seeböhm was badly wounded and died soon afterwards, but most of his records fell

4

into British hands, along with the survivors of his unit, captured while they tried to destroy as much information as possible.

This put the boot squarely on the other foot. It showed the British how much Rommel's ability to produce the unexpected depended on his being forewarned of their tactics by their own poor wireless security —and it also revealed how the Germans had been reading American diplomatic reports sent back from Cairo to the United States, keeping them informed of British moves. Both leaks could now be stopped—or the channels could be used to feed the Germans false information. The Germans knew that Seeböhm and his unit had been taken, but they had no idea how successful their men had been in destroying the secrets of their work. So long as they spotted nothing different in the content and frequency of British radio signals, they might conclude all was well— and go on acting on the basis of the information they intercepted.

Alamein, for Rommel, was but a staging post on the road to Cairo— the next step was to drive the British from their positions and set them in retreat once again. Although British strength was increasing by leaps and bounds, the best way to prepare for a real counter-offensive was to persuade them to attack in an area where the defences could be concentrated, and the geography be against them. Already Montgomery's Chief of Staff, General de Guingand, had seen a set of captured German maps of the area which showed that the Afrika Korps knew nothing of the ground conditions in an area of the front called the Ragil, where the sand was too soft to support heavy vehicles. Could they be persuaded to attack there? Messages were sent back to the Germans through a Cairo agent, codenamed Kondor, who had been captured by the British without his masters knowing. These signals said that the British would make their final stand at Alam Halfa, but that for the time being the defences were only half prepared. Later messages gave the British order of battle in the area, based on fact but altered to show the defences as far weaker than they really were. Finally, British combat maps which showed the Ragil area as 'hard going', ideal for tanks, were planted on the Germans. A British officer who had been arrested for unwittingly having allowed information to be passed to the Germans was sent into no-man's-land in a scout car carrying copies of the doctored maps. His car was blown up and he was killed, but a German patrol found the maps by his body.

So when Rommel's attack began on the first day of September, 1942, it followed the line the British themselves had chosen. Mines had been laid along their line of advance, air strikes harassed their columns, and when Rommel finally swung northwards to outflank the unexpectedly strong resistance, he ran straight into a double trap: his armoured vehicles bogged down in the sands of the Ragil and provided sitting targets for three armoured divisions dug in along the crest of the Alam Halfa ridge.

It proved a catastrophe for the Germans and after three days of furious fighting they began to fall back. Their casualties were three times those of the British. From now on, each day that passed would tilt the balance further against the Afrika Korps.

The vast desert battlefield was now narrowing to the strip of hard sand and low ridges which formed a bottleneck between the Mediterranean and the salt-marshes of the low-lying Qattara Depression. Here at last the growing fatigue and the dwindling fuel reserves of the Germans and the strength of the British defensive positions reached a balance, and the long retreat came to an end. Both armies dug in deeply, with solid defence lines massed behind belts of minefields.

Nothing could disguise the fact that the attack had to come somewhere along the thirty-mile strip between sea and salt-marsh. Could the enemy be deceived as to where and when along the narrow front the blow would fall?

Lightfoot was the code name for the Alamein attack and Operation Bertram for the deception plan. The first stage was to conceal the build-up of stores at the northern end of the front, which would otherwise give away the impending attack. Petrol cans were hidden in a network of old slit trenches, after careful checks had been made to ensure that no difference was made to their appearance from the air. Over three nights, three thousand tons of petrol had been brought in and hidden without any outward clue to its presence.

The remainder of the stores essential to the attack and the breakout could not be hidden so easily, but they could be disguised as something else. Piles of stores were covered with awnings which made them look like parked 10-ton trucks. Large concentrations of trucks might conceivably alert the Germans, but they were not in themselves evidence of a coming attack — as time passed and the trucks remained where they were, German anxieties decreased, and the hidden stores were accepted as part of the normal landscape of the opposing lines.

It was the presence of tanks and artillery which would really give the game away. So the huge armoured forces being built up for the battle were kept further back in staging areas with names like Murrayfield and Melting Pot, in the centre of the front but on a network of tracks which led towards the south. Having established in the Germans' minds that concentrations of trucks were harmless enough, more concentrations of dummy lorries were built up in the north. Then tanks could move forward during the night, to be hidden in their new positions by daybreak, under awnings which had already been identified as trucks. To prevent the Germans noticing that the tanks had moved, their tracks were swept clean and dummy tanks were left in their places. In the eyes of air reconnaissance nothing had changed: the armour was still well back from the front, massed ready for a move to the south.

To balance the lorries at the northern end of the front, used to conceal the armour and stores, the area at the southern end of the front was littered with huge dummy supply dumps. Every kind of coded wireless traffic, sent from transmitters which the German direction-finders would pin-point at the southern end of the lines, also pointed to activity in the southern sector. But was the evidence too perfect, too consistent? Without a flaw somewhere in the picture, might the Germans not suspect a deliberate attempt at deception?

So the southern staging area contained a double bluff: three dummy regiments of field artillery were sited in positions which were improperly camouflaged, so that the Germans could clearly see they were dummies. Then, as the attack grew imminent, real guns were moved into the camouflaged positions to open fire on the enemy—a deception deliberately revealed to distract the Germans' attention from others which were still hidden.

Ultra (see p. 50) had already shown that the Germans had no inkling of the attack in the north before the blow actually fell. Now it succeeded in deciphering signals which told of the sailing of the supply tankers which Rommel so desperately needed. A steady succession were due to leave from Italian ports, tempting targets for the Malta-based squadrons of the RAF.

This posed a very difficult problem for the British commanders. Sinking one ship could be explained by the enemy as bad luck, and sinking two could be a coincidence. But if all five tankers now making ready to sail were intercepted and sunk, then the Germans must conclude there was some massive security leak within their own organization, and the Enigma signals would be a prime suspect. Was the pressing need to cut off Rommel's supplies enough justification to risk the cutting off of this priceless flow of information?

In the end it was decided to go ahead. All five tankers were finally sunk in heavy sea mist which made the idea of chance encounters impossible to believe. Then Group-Captain Winterbotham came up with a brilliant stratagem. A signal was sent, in a cipher which was known to have been cracked by the Germans, congratulating a totally non-existent group of agents in Italy on the information which had resulted in the sinking of the tankers. Within days, more Ultra intercepts showed that the message had been read, and the story accepted as the truth. German units in mainland Italy were searching for the agents, and for the time being the threat to Ultra had been averted.

The Germans were in an impossible situation. Fuel was now desperately short, and the placing of their Panzer forces was crucial to the whole defence scheme. Long approach marches to meet an attack at the other end of the line would mean wasting petrol and risking breakdowns, as well as losing time. So the British deception plan resulted in the Panzers being split into two groups, one in the southern sector and one in the north.

Hiding the point where the attack would come was half the battle. Now the timing of the attack had to be concealed. Aware that the Germans thought the southern sector was the vital front, army engineers began building huge, fake water-supply pipelines, complete with distribution and pumping stations. There was only one possible explanation for all this effort—the water was needed to supply all the extra men being brought into the area for the main attack. Not only did this lend further weight to the idea of the attack coming in the south, but it was obvious from the rate at which the work was going ahead that it could not possibly be ready before the middle of November, 1942. No attack could therefore be mounted before then. In fact Montgomery was planning to strike the decisive blow at the other end of the front on 23 October.

So well did the deception work that Rommel himself risked a spell of home leave and hospital treatment, something he would never have done had he felt there was the slightest risk of an imminent attack. When it came, his subordinates were caught entirely off guard and reacted by swinging the Panzers northwards. They then withdrew German units which had been used to stiffen the Italian-held parts of the line. Montgomery saw his opportunity. As the initial attack began to lose momentum, he sent in the Australians, attacking along the coastal strip at the extreme northern end of the front, drawing the German defenders still further northwards. Meanwhile, at the weakest spot in the Axis front, where the German and Italian sectors met in the centre of their line, the enemy had been allowed to recognize the gun positions as dummies. In fact they had been secretly replaced by genuine artillery which, on 2 November, opened up with a furious barrage, heralding the final decisive blow. After two days' hard fighting, the breakthrough was complete; most of the Italians were cut off and the Germans were using the rest of their fuel reserves fleeing to the west.

The lessons of the desert war had already shown how powerful a weapon deception had become. After the end of the desert campaign, tactical deception gave way to strategic deception, its scale magnified a thousandfold.

CHAPTER THREE

THE WAR AT SEA

AT SEA, as on land, the war took time to develop. Partly the reason was that Admiral Doenitz's U-boat force was still far from full strength when war was declared, with only twenty-two submarines capable of coping with Atlantic conditions. Yet although only seven of these could be on patrol duty at any one time, in the first seven days of the war they sank 68,000 tons of Allied shipping, half the weekly loss which had come close to starving Britain into surrender in 1917. But the real cause was Hitler's so-called 'peace offensive'. Having had to fight for what he wanted in Poland, he was now determined to confirm its seizure by negotiation and to return to the conditions of peace which he had exploited so well. Heavy Allied losses at this stage would, he thought, prejudice his chances.

So orders went out, to commanders of submarines and surface raiders alike, to spare all passenger ships. Unfortunately the damage had already been done, for, in the tenth hour of the war, the *U30*, commanded by *Oberleutnant* Lemp, had attacked what he believed to be an armed merchant cruiser. These ships were converted passenger liners, armed with old six-inch guns and crewed by naval reservists, to serve as emergency convoy escorts; but Lemp's target was in fact the liner *Athenia*, which sank with the loss of 112 lives, twenty-eight of them civilian passengers from America.

This was political dynamite and Propaganda Minister Goebbels had to move quickly to defuse it. Grand Admiral Raeder, Commander-in-Chief of the German Navy, was made to issue an official denial that any U-boats had been in the area at the time, and Foreign Minister von Ribbentrop told the American Ambassador that no U-boat could possibly have been responsible. Goebbels himself insinuated that the true culprit had been a British submarine acting on Churchill's orders in an attempt to turn American opinion in favour of war with Germany.

When *U30* arrived back in Hamburg, Lemp narrowly avoided court-martial and was, with most of his crew, hurriedly transferred to the *U110*. The rest of the story was suppressed; the logbook of the *U30* had all entries for the date when the *Athenia* was sunk torn out, and Admiral Doenitz had all reference to the affair expunged from his personal war diary. To the world at large the *Athenia* sinking remained an unsolved mystery until Doenitz himself admitted the truth under cross-examination at Nuremburg.

There were other blows to German ambitions in the naval war. The loss of the pocket-battleship *Graf Spee* was a sharp reminder of the strength and professionalism of the Royal Navy. Beaten in an encounter with three smaller British cruisers, the German ship ran for shelter to the neutral port of Montevideo in Uruguay. It was vital for the British to keep her there until reinforcements arrived and made her escape impossible; but if they revealed this, the Germans might conclude that their only chance was to leave immediately. So began a skilful double-bluff. On the one hand, the British diplomatic staff in Montevideo kept up noisy official pressure on the Uruguayan authorities to force the *Graf Spee* to sail. At the same time the commander of the only British warship which could be seen from the land kept up a stream of signals to other ships, as if they were lying in wait just beyond the horizon. Gossip about the arrival of the *Ark Royal* and the *Renown* was dropped in waterfront bars; the Ambassador mentioned their imminent arrival to other diplomats in passing, and messages were sent to the two ships in an easily broken code as if they were already offshore.

For Langsdorff, captain of the *Graf Spee*, this was the last straw. Unable to face sailing out to certain and pointless death with his young and devoted crew, he scuttled his ship in the entrance to the harbour as his 72-hour time-limit expired. The *Graf Spee* blew up and sank in flames; her crew were taken prisoner and her captain shot himself.

After this disaster Hitler began to lose hope for his surface ships. He even renamed the *Deutschland* the *Lützow*, for he realized only too well the moral effect on the German people of the loss of a ship bearing their country's name. From now on the German war at sea would depend far more on the U-boat fleet and on their formidable armed merchant raiders.

The German Navy's biggest challenge came when Hitler decided to protect his northern flank by invading Norway in April, 1940. The crucial part of the plan was the sea crossing into southern Norway. If the troop transports and supply ships were attacked by a strong British naval force, there was little the *Kriegsmarine* could do to defend them.

By this time the German *B-Dienst*, the codebreaking and monitoring service, had intercepted British signals which told them that the British were planning to seize the port of Narvik in the far north of Norway. So the Germans concocted a deception plan which used the British invasion of Narvik as the key to their own objectives. Once their monitoring services showed that the British convoys were at sea, the Germans would send a strong naval force north to menace it. This, they knew, would draw off every British warship in the area to defend their invasion force and the seas off southern Norway would be clear for the Germans' own invasion fleet.

The plan worked beautifully. At the end of March, 1940, *B-Dienst* monitors picked up signals showing that the British force was assembling, and on 2 April Hitler set the invasion date for 9 April. The threatening force of battlecruisers and destroyers sailed immediately, and the British reacted as expected, the battleships and carriers of the Home Fleet withdrawing northwards with their escorting cruisers, leaving the main German forces to land in southern Norway almost without opposition.

The results achieved by the commerce raiders were disappointing. The introduction of heavily escorted convoys in the North Atlantic robbed the pocket-battleships of their most tempting targets — fast modern cargo ships sailing alone. An attempt to build up a striking force large enough to take on any likely opposition ended in disaster when Germany's newest and most powerful battleship, *Bismarck*, was caught and sunk while running for sanctuary in the French port of Brest.

This left the rest of the German force, the battlecruisers *Scharnhorst* and *Gneisenau* and the heavy cruiser *Prinz Eugen*, out on a limb in their French refuge. Brest was the target for frequent heavy raids by the RAF. Hitler, suffering from an acute attack of intuition, decided that something decisive was about to happen in Norway and, to guard against the possibility of Allied landings, thought the best place to assemble his last heavy ships was in the fjords, where training and refitting for operations with the *Bismarck*'s new sister-ship, the *Tirpitz*, could take place, But, he insisted, there was to be no long passage back through the Atlantic and the Denmark Strait, open to attack from British capital ships anywhere along the route. This time the German battleships would take the shortest and quickest route home, through the English Channel.

Hitler's naval advisers, faced with this directive on 12 January, 1942, were horrified; taking heavy ships through narrow confined waters under the gaze of the enemy's own coast-line spelled suicide. But the Führer was adamant; all he allowed his subordinates to do was work out the details by which his orders could be put into action. Tagged Operation Cerberus, the plan was drawn up by Vice-Admiral Ciliax, who was to command. Throughout the next four weeks no efforts were spared to mislead the British about what was being planned.

The basis of Ciliax's plan was that the British *did* expect the German ships to make a dash up-Channel, but he hoped to deceive them over the date and time. Reasoning that they would expect him to use the cover of darkness for running the gauntlet of the Dover Straits, this would mean a daylight departure from Brest, so that the cat would all too soon be out of the bag. By leaving Brest at night, however, he could make his ships' departure seem like a normal training exercise. His real intention might not then be apparent until the following day. At the same time careful time-tables were drawn up for heavy air escorts and fleets of escorting destroyers and patrol boats.

Before the force could sail, a safe route had to be swept through the minefields along the Channel coast. But any increase in minesweeping activity would soon reach the eyes and ears of the French Resistance, who would report it to London where the obvious deductions would be made. So Admiral Ruge, commander of the minesweeping forces, drew up a complicated plan whereby, without any apparent increase in activity, the whole channel was swept length by length, the last gaps being covered only days before the operation began. Even then, things were close; to prevent the French noticing that reserve stocks of marker buoys were being taken from the harbours, Ruge stationed anchored minesweepers along the channel instead.

In Brest the plans continued under the strictest security. Because everything was put into operation locally, there was no coded radio traffic to attract attention, and the Germans were clever enough to pretend to be secretive while at the same time making sure that the French dock-workers and civilians saw exactly what they wanted them to see. Ciliax ordered sun-helmets and tropical clothing from French suppliers to be delivered to the ships; drums of lubricating oil of high viscosity for use in hot climates, prominently labelled '*lubrificants coloniaux*', were left lying on the dockside.

The departure of the ships, on the night of 11 February, 1942, could not possibly be concealed from the French. But the story was spread that they were going out on fleet exercises and gunnery trials in the Bay of Biscay; the French harbour administration was instructed to have target-towing vessels ready in the exercise area, and also to have tugs and net-laying vessels ready for the ships' return to port the following day. Many of the senior officers of the fleet accepted invitations to a hunting party at Rambouillet on the 12th, and even the crews, outside the select few in the know, confidently believed they would be back in Brest early on that day.

Instead, once free of the harbour, after delays caused by a British air raid and by *Scharnhorst* hitting a torpedo net, the force turned north and then east, escorted by six destroyers. At daybreak the following morning the escort was backed up by fifteen large torpedo boats and twice that number of fast patrol-boats. Heavy air cover was provided by relays of fighters from coastal airfields throughout northern France.

Having themselves suffered the catastrophic loss of the *Prince of Wales* and the *Repulse* by Japanese air attack only weeks before, many of the British senior commanders had decided that the Germans would never dare bring their heavy ships so close to British bomber bases. Those who still believed in the danger took what steps they could; Commander (later Vice-Admiral Sir Norman) Denning of the Operational Intelligence Centre of the Admiralty persuaded the Flag Officer, Submarines, Admiral Max Horton, to station HMS *Sealion* in the approaches to Brest. But, because of their delayed departure, *Sealion* had withdrawn from her station to deeper water, and an RAF patrol plane also watching the harbour approaches had to return to base with a defective radar set. It eventually returned to its patrol station, but by then the Germans had gone, and by incredible good luck, the aircraft covering the next sector also suffered a defective radar set.

The result was that the Germans were only spotted by an RAF Spitfire at 1035 on the morning of the 12th, by which time they were off Le Touquet and approaching the Straits of Dover. Most of the motor torpedo-boats assigned to watch the narrows had been dispersed and all their tactics and training had been worked out on the basis of a night action. Because it was daylight, the German ships were steaming at full speed, the remaining MTBs were in full view, and their attacks were fruitless. Later attacks by bombers, destroyers and torpedo bombers all failed. The only British weapons which inflicted any damage were mines which had escaped Ruge's sweep, two of them being set off by *Scharnhorst* and one by *Gneisenau*; but all the German ships reached their destination safely in an operation ending in complete psychological victory for the Germans.

Whatever its psychological success, however, Operation Cerberus was the swansong of the German surface forces in terms of actual operations. For the rest of the war their role was limited to that of a deterrent.

After the sinking of the *Bismarck* in May, 1941, the navy decided to curb the surface raiders by attacking their one Achilles heel—the network of supply ships on which they were dependent for fuel, food and ammunition. Working on the basis that all the supply ships available to the Germans would have been mobilized for the *Bismarck* operation, the Home Fleet began a concerted search which ended in mid-June, 1941, with the loss of six tankers and three supply ships.

After this the menace of the surface raiders was spent. In any case their sinkings had been small compared with the U-boats. By that time their record totalled eighty-six ships, just over half a million tons—but their psychological effect had been out of all proportion to their actions. In particular the codes they had helped to break were to give their submarine colleagues a powerful additional weapon which almost won them the Battle of the Atlantic.

For it was in these bleak northern waters that the sea war was about to take its most decisive turn, on the direct supply routes between the United States and Britain. Here, above all, was Germany's great opportunity. For all that the RAF might deny the *Luftwaffe* command of the skies over England, for all the achievements of the Eighth Army against Rommel in North Africa, victory here would mean England's certain strangulation and inevitable surrender.

The vessel which nearly brought this triumph about seemed puny enough—the typical ocean-going Type VII U-boat, only 220 feet long and displacing 800 tons. She carried fourteen torpedoes. While on the surface her best speed was seventeen knots and, by cruising on the surface, her diesel engines could keep her running for 6,500 miles. But once she was forced to submerge her speed and endurance dwindled to almost nothing. She was dependent on battery-powered electric motors, which allowed a top speed of $7\frac{1}{2}$ knots, slower than most of the convoys and all the escort vessels. At this speed she could keep running for a matter of hours at the most before being compelled to surface, there to run the diesels to recharge the flattened batteries. Only by slowing literally to walking pace could she conserve her batteries to stretch the underwater range to eighty miles and her endurance to the best part of the day.

All this meant that the U-boats were acting as surface warships for most of the time. Diving was an emergency measure carried out as a last resort. So if the British could merely force them beneath the surface and keep them there for long enough, the submarine force would be paralysed. Lone U-boats could be kept down long enough for the convoy they were chasing to disappear over the horizon; then the escort vessels could pull away at high speed, leaving the submarine helpless to follow.

So U-boat tactics had to be carefully planned and co-ordinated. Convoys had to be found in the vastnesses of the ocean, and as many U-boats as possible assembled to make surprise attacks at the vital moment, to overwhelm the escorts and sink as many merchantmen as possible before the convoy could escape. And the key to success, for attackers and defenders alike, was radio—both for communications and for eavesdropping on one another's dispositions and intentions.

Other information was valuable in all kinds of ways. By careful monitoring of the Germans' coded signals, a great deal of information could be built up without knowing what the messages actually said. By using an increasingly sophisticated direction-finding network which would pin-point the position of the submarine which sent each message to within a few miles, careful recording, cross-checking and comparison could enable Allied intelligence experts to produce accurate guesses as to the subject of each signal. Short messages were usually convoy sighting reports, longer ones gave weather data, reports of actions or damage and the longest were usually detailed end-of-cruise summaries. Tape recordings of each message helped monitors to recognize the 'fist'—the individual Morse rhythm—of a particular operator, which could help to identify a particular submarine.

Snippets of background information on the U-boat service as a whole —harbour gossip, numbers of individual submarines sailing on patrol at a particular time, which could then be tied in with specific intercepted signals from the appropriate positions—were all sent back to England by agents inside occupied Europe, working in the neighbourhood of the submarine bases themselves.

This information was useful for another psychological reason. Although many U-boats were sunk with their entire crew when depth-charged beneath the surface, others managed to stay afloat long enough for some at least of the crew to escape. When these U-boat men arrived at the prisoner-of-war reception centres they were met by men who knew almost as much about the service of which they had been a part as they did themselves. Their interrogators knew so much about conditions at the submarine bases, about German tactics and attacking methods, about new techniques and new weapons and about the names and reputations of commanders, that it became psychologically very hard for the prisoners to resist the impression that their questioners knew everything already. If all they did was reveal the standard name, rank and number, this obviously detailed knowledge on the part of their interrogator was disconcerting enough, but if they let slip the number of their U-boat, the information the questioner had on his records proved even more frighteningly detailed. Soon the careful questioning of prisoners was revealing ever more useful and up-to-date information, and once the severely shaken prisoners were released from questioning to join their comrades in the camps, bugged rooms were able to collect even more information when they proudly revealed to the others the information they had been able to keep from the British!

The Germans had a trump card in their mastery of the Allied naval codes, which told them exactly when and where to look for approaching convoys, but they lost their advantage by forgetting one cardinal rule— always try to conceal your advantage from the enemy. Careful analysis in the Admiralty's submarine tracking room showed that whenever a convoy's course was altered to take it clear of a known U-boat patrol line, after a short delay the positions of the U-boats would begin to shift, until eventually they lay across the convoy's new line of advance.

Sometimes a fast ship would be allowed to sail alone, without convoy protection, relying on a speed faster than any U-boat; but time and again a submarine on patrol near the ship's track would blunder into her. Both ship and convoy would be maintaining wireless silence, so there was no question of the U-boats homing in on their signals. The German movements were designed to be as casual as possible, but the conclusion was inescapable: British ciphers were being broken and signals read by the enemy; the long and complicated business of replacing them with new ones was imperative. The system itself could not be changed until all commanders had copies of the new ciphers, but by the middle of 1943 British communications were once again a closed book to the Germans.

In some respects a U-boat was most vulnerable in home waters. Tired and worn out at the end of a long cruise, or nervous and out of practice returning to sea after a spell of leave, these were the most dangerous periods. The U-boat Command tried desperately to find some answer to the deadly radar-guided Allied air patrols over the Bay of Biscay before losses climbed and morale slumped. The British were already working on a balloon which, when coated by a metallizing process developed from that used to make the artifical jewellery worn by Vivien Leigh in the film *Caesar and Cleopatra*, produced a very strong radar echo capable of misleading the enemy. The German equivalent was called Aphrodite, and was much more successful against the Allied high-definition radar. When the Director of Naval Intelligence, Admiral Godfrey, realized this, he persuaded British radio stations to mount a strong anti-Aphrodite campaign. The Germans drew the conclusions he hoped they would draw—that the Admiralty were worried by the German experiments; so they persisted in investigating what Godfrey knew to be a complete blind alley. In the same way, another carefully planted rumour that the British aircraft were homing on the infra-red radiation from the U-boats resulted in the Germans pinning false hopes on developing a special paint for submarines which could reduce these emissions.

When it came, the end came quickly. After topping the pinnacle of success in March, 1943, when they sank half a million tons of Allied shipping in just three weeks, the U-boats plunged into the abyss of defeat. During the months of April and May they sank twenty-three merchant vessels, but escorts and aircraft accounted for no less than twenty German submarines. This ratio was too much. On 24 May Doenitz ordered all his U-boats to withdraw from the mid-Atlantic convoy routes.

From now on the submarines would have to look further afield for their targets. Aided by 'milch-cows'—large submarines built to act as refuelling tankers—and supply vessels to keep the smaller attack submarines replenished and rearmed when patrolling far from home waters, they could reach the Caribbean, the South Atlantic and the Indian Ocean. But their days were numbered, even in these distant waters, thanks to a pre-war British intelligence coup. A report had been received from the Polish Secret Service that a Polish technician had been repatriated from Germany where he had been working at the Heimsoeth code-machine factory in Berlin on a battery-powered encoding and decoding machine for the German Armed Forces. Although the Pole had been limited, for security reasons, to working on a single part of the machine, he had taken enough interest in what his colleagues had been doing to be able to make a fairly accurate reconstruction of the whole unit.

The British Intelligence network, who were sent the message by their Polish opposite numbers, acted quickly and the young technician was sent out of Poland through Danzig to Paris, where he built a wooden mock-up of the machine on which he had been working.

Although incapable of cracking German codes on its own, the reconstruction was promising enough for the Poles to be asked for more. Working through a local Swedish-run intelligence network, they managed to produce a complete machine which was flown to England only days before the war began. Originally an enciphering and deciphering machine, it was invented by a Dutchman, Hugo Koch of Delft, in

1919. He assigned the patents to a German engineer, Artur Scherbius of Berlin, who built the machine from Koch's plans. He called it the Enigma, after Elgar's *Enigma Variations*, and it was exhibited at the 1923 Congress of the International Postal Union. Then Scherbius sold the patents to yet another company. Finally the resurgent *Wehrmacht* came across the Enigma machine, modified it and improved it for their own purposes.

The whole principle of the machine hinged on the fact that, even with an identical machine, anyone who intercepted the messages could not decipher them without knowing the keys, or settings, of the code-wheels and connections within the machines. Although the French had built a replica of the machine before the war, they were dependent on sources within Germany for the vital keys to unlock the messages.

This was the reason why Hitler allowed the machines to be used in places (army units, submarines, etc) where they might be captured. Only if the enemy knew the keys was there any danger of the messages being read. There were different keys in use for each type of communication, and in many cases the keys were changed each day and night. But the British evolved new techniques of statistical analysis which allowed the Post Office experimental computer to try out whole ranges of possible keys in minutes, searching for any significant patterns which would help narrow down the search and finally identify the key actually in use. But this took time, and demanded a number of messages with the same key pattern, so that not all signals could be completely cracked straight away.

Early in 1940 a special unit was set up at Bletchley under Group-Captain Winterbotham, head of the Air Section of the Secret Intelligence Service, to decode all the Enigma messages picked up by the radio monitoring services and generally to act as a Shadow OKW (Wehrmacht Headquarters) in issuing copies of all the genuine OKW's orders and dispatches which were picked up and deciphered. Great care had to be taken to avoid this goose being killed before it had produced its quota of golden eggs. Whenever information came through the Enigma code-breaking operation, it was to be treated as super-top-secret (hence its code-name 'Ultra') and revealed only to those who were responsible for initiating action as a result of the knowledge acquired. Secondly, there had always to be another explanation for these actions, to guard against the Germans deducing that their codes had been broken.

Cracking the German Naval ciphers took a great deal longer because their version of Enigma had two more code-wheels than the standard version. But real progress had been made by early 1943, at a time when Allied merchant losses were approaching their zenith. The temptation to use information provided by the Ultra intercepts to re-route convoys and escort forces was enormous. When the information could also have been provided by direction-finding this was done, but before long Ultra messages showed that Doenitz himself was actively investigating security leakages.

Fortunately, the extra complexity of their own Enigma system led the *Kriegsmarine* investigators to suspect other sources: either British agents in the ports, or first-class direction-finding. But Ultra showed the British that the Germans *were* reading their naval and convoy codes, so that changing them became top priority. It took time for all ships and bases to be issued with new cipher books and key tables, but, once that was done, the U-boats were cut off from a vital source of intelligence.

By May, 1943, however, Allied losses had reached a level so serious that Churchill decided to allow Ultra to be used to the full against the U-boats, without cover. Ironically, because by this time the Germans had suspected and then cleared their own Enigma system as a possible leaking source, the U-boats went down to sudden and final defeat without Doenitz ever realizing why.

Now Ultra was producing information which was vital to this new phase of the naval war—nothing less than messages arranging the rendezvous between cruising U-boats in distant waters and their supply ships. By sinking these, the rout of the U-boats in the central Atlantic could be matched by their defeat in more distant waters.

There was inter-Allied disagreement about the action to take over the supply tankers. The Americans wanted them all sunk as quickly as possible, but Air-Marshal Sir John Slessor, Commander-in-Chief of Coastal Command, managed to persuade them to adopt a more gradual programme. It was vital to make the Germans think they were losing their U-tankers through bad luck and superior Allied weaponry rather than pushing them into changing their codes and cutting off all future supplies of information to Ultra.

Fortunately for Allied intelligence, Slessor's approach worked. One by one the milch-cows were picked off at their rendezvous points. Yet unlike the Allies, who had realized that the Germans must have been reading their coded signals, the Germans suspected nothing. Doenitz's headquarters had six full-time officers watching for any sign of Allied tactics or movements which could not be explained away by what they already knew was a brilliantly effective system of radar and high frequency direction-finding stations. So carefully was the Ultra information acted upon that they found none. Doenitz himself wrote, 'Except for two or three doubtful cases, British conclusions are based on data regarding U-boats which are readily available to them, on U-boat positions and on their own plotting of the boats' movements, combined with a quite feasible process of logical deduction'.

So the German cipher system remained unchanged, to the immeasurable benefit of Allied operations, as we shall see in later chapters. The Germans decided they had lost the Atlantic battle to superior Allied weaponry, and any turning of the tables would have to wait until their own new weapons, the schnorkel, the homing torpedo, the rocket and the hydrogen peroxide submarine, were ready to come into action.

By 1945 there were dozens of streamlined super-submarines under construction in the yards at Hamburg and Bremen. They were being batch-built in complete sections, the different sections being joined together by welding on the slipways just before launching.

These were formidable craft indeed. Because of the combination of high under-water speed, huge storage batteries and a schnorkel breathing tube which allowed the submarine to run on its diesels just below the surface, they were almost immune from air attack. They were faster than many of the escorts, but the fact that they were compelled, on account of Allied command of the air and sea, to stay submerged for most of the time, meant that the Allied monitoring services had no information at all on the new U-boats' capabilities. They could even fire torpedoes, on instruments alone, from depths of 150 feet, secure from radar detection, before making off at high speed underwater. Enough of these boats could transform the undersea war, even at this late hour.

There was one hope of restricting the operations of the new U-boats. Because Germany was now totally on the defensive, and because the bases in France had long ago been lost, the new boats were sent into action from pens in Germany and Norway against the convoys in British coastal waters. Some areas were protected by minefields, which kept the U-boats away without the need for escort vessels or patrols at that point. The new submarines were far too valuable to risk their destruction by known hazards like this. So if the German Naval command could be persuaded that there were many more minefields than there actually were, whole areas of coastal waters would be denied to the new vessels. With a smaller area of sea in which the new submarines could be found, escort patrols could be strengthened to keep up a more effective watch.

So it was that one of the last of the wartime naval deceptions began. This involved a German double-agent, Hans Schmidt, code-named 'Tate', who had already played his part in other deception programmes by sending the Germans false information which had been fed to him by his British masters. This time his cover story allowed him access to confidential Royal Navy reports, dealing with submarine sinkings. As details of each sinking came in from the escort forces (these were, of course, the older and more conventional submarines), he was allowed to radio his German contacts claiming that such-and-such a U-boat had been sunk in such-and-such an area, by *hitting a mine*. All the Germans themselves knew was that, several days after receiving the message from Tate, their own records would confirm that nothing more had been heard from that particular U-boat. So they would then pencil in another section of their maps as an Allied minefield — or so the Allies hoped.

In fact the operation worked beautifully, but what clinched the agent's messages in the eyes of the Germans was a stroke of almost unbelievable good luck. Tate's deception depended on the fact that none of the U-boats now made the regular reports to headquarters that had been routine two years before, and most certainly no U-boat could report back with details of how it had been sunk, to give the lie to his minefields story. On the other hand, the fact that his story could neither be denied nor corroborated may well have given the *Kriegsmarine* reason for doubt, as the reports continued to flood in, and more and more waters were being denied to their new submarines.

But one U-boat *did* get into trouble. It was badly damaged by a sudden explosion, but, by blowing all tanks, managed to reach the surface long enough for the crew to escape and scuttle the submarine. Before she went down on her final dive, the crew radioed headquarters, having nothing to lose by giving away their whereabouts. They reported that they had been sunk by a mine, and gave her final position before signing off. In fact, the mine had been a drifting one, although there was no way they could have known it. But, wherever it had drifted from, the position in which the U-boat sank was right in the middle of one of Tate's imaginary minefields. After that the German High Command had no doubts whatsoever. Strict orders went out to U-boat commanders which, in the end, virtually closed off 3,600 square miles of sea to the German Navy's submarines. And with them went the *Kriegsmarine*'s last hope of avoiding total defeat in the war at sea.

CHAPTER FOUR

WAR IN THE AIR

As WITH the war on land, the air war burned on a slow fuse for the first months. The Germans had their hands full in the East; of the *Luftwaffe*'s 3,600 frontline fighters and bombers, no less than 2,600 were ranged against the unfailingly courageous but totally outnumbered squadrons of the Polish Air Force. But Allied Intelligence had no way of knowing this directly, especially since German policy leading up to the outbreak of war had been calculated to produce exactly the opposite impression. Goering's top *Luftwaffe* aide, General Karl Bodenschatz, had revealed to Paul Stehlin, later Chief of Staff of the French *Armée de l'Air*, details of how the *Luftwaffe* was being reorganized after Munich to treble its former size and striking power. By the spring of 1939 the British and French Air Force staffs had formed the most alarming impression of the *Luftwaffe*'s muscle. One estimate said that Goering had enough bombers to send a thousand-bomber raid against London every day for a fortnight.

The swift aerial subjugation of Poland did nothing to allay these fears. After the front-line squadrons of Messerschmitt 109s had cut the defending Polish fighters to pieces in a matter of hours, the skies were cleared for phalanxes of dive-bombers. The cranked-wing Stukas proved an appallingly effective psychological weapon against ground troops with no air support. Their unmistakable appearance and the method of their attack, which consisted of a half roll followed by a near vertical dive straight at the target under full power, had a terrible impact on those experiencing it for the first time. Everything was calculated to inspire terror. In fact extended dive brakes kept the speed down, but the bombers' falling flight was made to seem faster by the use of sirens to produce a shrill wailing scream. Dive bombers as a weapon were a poor substitute for artillery; they could be brought into action quickly but they were far less accurate. Each plane had only the same chance of hitting the target as the one before it or the one after, as there was no way of correcting the fall of shot. They were vulnerable to anti-aircraft fire, and even more so to fighter attack—but no one realized this at the time.

The results of all this propaganda, backed up by set-piece demonstrations from Guernica in the Spanish Civil War to the bombing of
Warsaw, were to produce an uneasy half-truce in the air. People and
politicians all over Europe were convinced that the bomber must
always get through, that war would produce a holocaust of burning
cities and populations decimated by air-borne gas attacks within hours
of hostilities breaking out, and faced with this prospect, they quailed at
the idea of unleashing their own bombers against the enemy.

The Germans were equally worried about the RAF's Bomber
Command; they knew the vital Ruhr industrial region lay well within
its range, and they credited Britain's bomber force with far more
effectiveness than it possessed at the time. So German bombers refrained
from raiding the west, while the British and the French behaved as if
dropping explosives on German soil would be breaking the law.

But once Germany had won the land war in the west the struggle in
the air gained a new and decisive importance. Before they could bring
their apparently invincible army into action on British soil, the Germans
had to smash the RAF's defending fighter force.

The struggle hinged on two technical developments – radar and
codebreaking. But any step forward in these fields can easily be countered *as soon as the enemy realizes what is happening*. Only if the truth
can be hidden or distorted can tactical advantages like these be turned
into campaign-winning weapons. Radar, for example, was a priceless aid
to intercepting German raids – but the network of sites strung out along
the southern and eastern coasts of Britain on which this warning
depended were themselves pathetically vulnerable to German attack.
They looked innocuous enough to outsiders' eyes, but would the
Germans realize their significance?

The battle proper started at the beginning of August, 1940. In a
brave attempt to keep the Germans off the scent as far as the RAF's
dependence on radar was concerned, much was made of the efforts of
the Royal Observer Corps who manned isolated observation posts
armed only with binoculars, height-finders and telephones. In fact,
valuable as their work undoubtedly was, Fighter Command needed
earlier warning of enemy attacks if it was to scramble squadrons in
time for them to reach a height where they could meet the enemy on
equal terms and scatter them before they could unload their bombs.
With many of the most vital radar sites and fighter airfields within
minutes of the coast, time was to be the deciding factor.

Ultra, too, was delivering invaluable information on roughly where
and when the main German raids would come each day. No detailed

information on targets was given, since these were sent to units by means other than radio. But, to take one example, on 15 August, 1940, the RAF did at least know that attacks could be expected to continue all day. Even this much information was vital, since it made sure that an adequate reserve of refuelled and rearmed fighters was kept ready, come what may, to deal with new waves of attacks.

Somehow each attack was met by some defending fighters, no matter how few, or how hurriedly they had been assembled. Each bomber formation was put off its aim by resolute attacks, which not only reduced the resulting damage to a fraction of what it might otherwise have been, but also wreaked havoc on the morale of the *Luftwaffe* aircrew. Before long they found themselves expecting to be intercepted and attacked on every flight. The continual presence of defending fighters hinted at much larger numbers than the RAF in fact possessed, and since German estimates at the start of the battle were not far from the truth, the feeling grew that their efforts to beat the RAF into submission were having no effect at all on its actual strength.

Then, at the very moment when their attacks on British fighter airfields were indeed beginning to cripple Fighter Command, the psychological impact on an otherwise ineffectual (in material terms) RAF raid on Berlin caused the Germans to switch the *Luftwaffe* bombing effort to London instead.

This decided the outcome of the Battle of Britain. The bombers wreaked havoc on London's streets and docks, but as a target the capital was better able to absorb punishment than the airfields, and was far less decisive in terms of the battle then being fought. Fighter Command used the respite to repair the airstrips, and by the third week in September the *Luftwaffe*, though still attacking bravely, admitted defeat. In occupied Holland the Germans began ordering their parachute troops to stand down from instant readinesss. The invasion threat was over.

But another threat still remained. Goering had found, as had Bomber Command before him, that the daylight sky full of defending fighters was no place for even the most modern bombers. Again, like Bomber Command, the *Luftwaffe* switched from the daylight blitz to night bombing; and, again like Bomber Command, the *Luftwaffe* was to find the friendly cloak of darkness an aid to the enemy as well. How could they hope to find their targets in blacked-out Britian, even by the most impeccable navigation?

In fact the *Luftwaffe* had already shown the difficulty of target finding, even in daylight. A force of Heinkel 111 medium bombers took off from an airfield near Munich to attack the French fighter field at Dijon on 10 May, 1940. It was a warm afternoon with plenty of cloud and the German aircraft lost their way, but, emerging once again into clearer weather, they found themselves near one of their alternate targets, the airfield on the outskirts of the town of Dôle, thirty miles to the south-east of Dijon. They lined up on their bombing runs, each dropped a stick of bombs too hurriedly to be sure of their aim, and vanished back into the clouds to start the flight back to their base.

But their navigation had been disastrously wrong. They were more than a hundred miles north-east of where they thought they were, and the town which passed so briefly beneath their bomb-sights was not Dôle at all, but the German town of Freiburg in the valley of the Rhine, which also had an airfield on its outskirts. Ten of the bombs fell on the airfield, but three times as many dropped in the town itself, eleven of them on the main railway station and two in a children's playground. Eleven German soldiers were killed as well as forty-six civilians, including thirteen women and twenty-two children.

By that evening the German authorities knew the truth. The Freiburg police had collected fragments of the bombs, some of which carried codes identifying them as German bombs delivered originally to that same Heinkel 111 bomber base near Munich. The truth, however, was not to leak out. Apart from the blow to civilian morale, not to mention the confidence of the bomber crews, who might be inclined to avoid bombing any target if they were not absolutely sure of its identity, it was too good a propaganda opportunity for Goebbels' Ministry of Public Enlightenment to let slip. If everyone believed it was a raid by the French, so much the better. Here were the enemy ignoring the civilized restriction which had so far governed the air war in the west, unleashing a torrent of high explosive on innocent German civilians.

So the propaganda blasts went out, trumpeting to the world the guilt of the French for attacking an open town in defiance of all the rules of war. Whatever the effect on neutral listeners, who could hardly have imagined the Germans would deliberately bomb their own citizens to make a propaganda point, there was the added bonus that the Germans themselves now had an excellent pretext for any offensive bombing of their own. The Allies, they could always say afterwards, had struck the first blow.

But the fact remained that bombing, especially night bombing, was still a disconcertingly difficult business. The Germans had already begun developing a whole series of blind navigation aids, using different combinations of radio beams, which could lead them accurately to their targets even in total darkness. Fortunately British Intelligence was aware of this. Back in November, 1939, a package of letters had been delivered to the British Embassy in Oslo, containing vital information on all kinds of projects from rockets to radio beams. The Oslo Letters, as they were called, mentioned a set of German research projects from long-range rockets to radar, but there was nothing to prove whether or not these things really existed. It could be a deliberate plant to encourage the Allies to worry about new weapons which never were, both to undermine morale and to lead them to waste time and resources looking for further evidence, which might of course lead them to miss tell-tale signs of real value in other fields. This was an obvious enough danger. Another was that the writer of the letter may have been sincere in his intentions, but may have been misled by incomplete knowledge or by German home-consumption propaganda into believing in non-existent technical advances.

Yet the information in the letters proved, in case after case, to be totally genuine. One German radio beam system was countered, not by jamming, but by bending the beams so that the bombs fell in fields instead of on crowded cities. But its successors proved more difficult to block. One tragic demonstration of the effectiveness of radio-guided bombing was given by the raid on Coventry on the night of 14–15 November, 1940. This was one case where the Germans did give away advance warning of their intentions. At 3 o'clock on the afternoon of the 14th a signal was picked up by the Ultra unit which mentioned the city by name as that night's main objective. Any attempts at mass evacuation would destroy Britain's most vital source of information immediately. The Germans would be bound to hear of any measures involving hundreds of thousands of people, and losing this priceless source of information would cost far more lives in the end than the raid could possibly claim. Anyway the warning was too short to undertake effective measures to alert everyone and move them out of the city. All that could be done was to alert the fire services, the police, the hospitals and ambulance services and the air-raid wardens and their teams. Decoy fires were prepared, to confuse the German aircrews as to which were the real objectives, and the jamming equipment, set up to counteract the newly discovered X-Apparatus, was made ready.

The result was a tragedy. Although the jammers should have worked, a technical hitch which was later corrected meant that the German pathfinder crews were able to identify the true signals and pinpoint their positions exactly.

The force numbered 550 bombers of which no less than 449 found the city, already burning from the incendiaries of the pathfinder crews. Between them they dropped 500 tons of bombs and showers of incendiaries. 380 people were killed, the cathedral and city centre destroyed, a dozen vital aircraft-component factories put out of action and a whole industrial area paralysed for more than a month by cutting off power supplies and gas mains.

Fortunately for Coventry, and for Britain, the *Luftwaffe* did not return to the city in time to follow up its real and psychological advantages. The defences of the city were overwhelmed, the fires were still burning, making it easy to find at night, and the population were still badly shaken from the first raid. The future looked bleak, especially as signals intercepted by Ultra went back to using indecipherable code-words to identify targets after this one lapse. The only bright spot in the gloom was that the electronic counter-measures experts had corrected their error and from the end of 1940 were able to jam the beams successfully.

After this the Germans reverted to visual navigation methods. Sometimes the glow of moonlight on water could give flyers an exact position, even far inland. When the Germans attacked the Liverpool docks for night after night, careful analysis of their track showed that they were making their final approach along a line of five reservoirs which linked the Trent Valley with the broad, easily-identifiable estuary of the Mersey. So the Admiralty's Department of Miscellaneous Weapons Development was put to work to make water less obvious, and Professor (later Sir Eric) Rideal's earlier experiments at camouflaging the course of a stream by spraying the water with a fine coating of coal-dust were adapted to covering up larger stretches of water. In fact, the experimenters found, it was easier to cover up reservoirs and lakes with floating camouflage netting, but at one point their technique worked so well that an old gentleman, out for a walk with his dog, walked right into the Coventry Canal, which had been experimentally treated with coal-dust, under the impression that it was a newly-surfaced road.

Covering up signs which might lead the bombers to their targets was one solution; another was to mislead them over the target itself. The Petroleum Warfare Department had already evolved a spectacular decoy system to be used during air raids on refineries or oil tanker terminals. The principle was to dig circular trenches at intervals, each trench being a foot across and lined with clay before being filled with a mixture of clinker and breeze-blocks, to make an inert filling which could then absorb oil to make an impressive but controlled blaze. When the bombers attacked the oil tanks, operators could set off the oil-filled trenches by electric detonators, which set off mixtures of magnesium and gun-powder which in turn set off the oil. The flames and smoke looked as if the plant was already ablaze, and explosions were simulated by petrol-filled lavatory cisterns, the chains of which were pulled by remote control. The idea was that the bombers would think their work was done and call off the attack, whereupon the fires in the oil-filled trenches could be allowed to die down harmlessly.

In North Africa, where the Italians were beginning to raid British forces in Egypt, Jasper Maskelyne was putting his specialized talents to brilliant use to confuse the enemy. Faced with the vulnerability of the British fleet in Alexandria harbour to night attack by bombers, Maskelyne decided that the only answer was to build an entirely new Alexandria a mile away, where a deserted cove called Mariut Bay resembled the shape of the harbour at Alex quite closely. There he set up a fleet of false ships, made from canvas, cardboard and wood, and his men toiled to lay miles and miles of wire for lights and other effects. Remotely-controlled explosive charges were also laid, along with flares, and as soon as the bombers were known to be on their way, Alexandria proper went into total blackout, while all the lights at Mariut Bay went on, fed from batteries of portable generators. As soon as the bombs began to fall, the explosives were set off, along with flares to represent fires and anti-aircraft gunfire. In the confusion all the bombers were attracted to the decoy site like flies around a jam-pot, while the fleet lying at anchor in the real harbour went completely unmolested.

The *Luftwaffe* and the Italian *Regia Aeronautica* had their problems, but so did Bomber Command. Despite the difficulties which enemy bombers were clearly finding in hitting targets in Britain, Bomber Command went on producing the most outrageously optimistic reports. By the autumn of 1940, some experts were cheerfully estimating that German morale was in ruins, and that a quarter of the country's productive capacity had been destroyed by British bombing. The truth was very different, and while the firm belief that every bomb loaded into the bay of a bomber bound for Germany was going to do real damage was essential to the morale of over-stretched aircrew and blitzed civilian alike, the obvious discrepancies between even the objective BBC communiqués and the lack of damage on the ground was an even better fillip for the Germans. Time after time Allied broadcasts would announce that targets in a particular city had been raided, whereas the German records showed that no bombs were recorded as falling anywhere in the area in question. Sometimes neutral correspondents, like the Americans, would be taken to visit these cities to show how untrue the British claims really were.

The simple fact was that the bomber crews were doing their best with the cruelly inadequate equipment at their disposal. Navigating by dead-reckoning, with the enormous errors inherent in this kind of flying at night, was no way to hit anything as precise as an individual city, let alone a pin-point target. The Germans too were experimenting with decoy fires and flares, set off in the immediate neighbourhood of a raid, as soon as the bombs began to fall. To a bomber crew, tired and confused after hours of flying through darkness, the temptation to drop their bombs on an area lit by fires and explosions was well-nigh irresistible. The result was that when the RAF submitted their whole collection of bombing-target photographs belonging to the Photographic Reconnais-sance Unit to expert examination in 1941, the figures showed that only about one aircraft in three had succeeded in getting within even five miles of its target, and in vital, heavily-defended areas like the Ruhr valley, this proportion dropped to only one in ten. Not only that, but Professor Zuckermann's experiments a few months later showed that British bombs were themselves far less effective than the German ones, to the point where it took five tons of bombs dropped on Germany to kill, on average, a single German civilian.

Knocking Germany out of the war by precision bombing of industrial targets was, on this basis, an utterly impossible assignment. Instead, Professor Lindemann, Churchill's principal scientific adviser, estimated that a concentrated attack on the morale of the German working population would be far more effective. There were fifty-eight towns and cities in Germany with populations of more than a hundred thousand. The areas of working-class suburbs were large enough in each case to provide the kind of target which could be hit with sufficient accuracy even on moonless nights, when the psychological effect would be greatest. By turning the whole of Bomber Command's resources on this objective, Lindemann estimated that from March, 1942, when his figures were drawn up, fifteen months should see a third of the German population bombed out of their homes, and morale collapsing as a result. Others, like Professors Blackett and Tizard, disagreed; but in the end it was Lindemann who prevailed, and the result was the so-called area offensive. In other words, the physical destruction of vital parts of German industry was to give way to the psychological destruction of a large part of its work force.

In the meantime the Germans were developing defences of their own. At the beginning of the war, they had air-defence radars like the long-distance *Freya* which could give enough warning of an approaching raid for night fighters to take off and climb to intercept the bombers. By the autumn of 1940 the shorter-range *Wurzburg* radar could give the height of the incoming bombers and could feed this information to the searchlights, which could then pick out individual targets for the patrolling night-fighters. But the really formidable system was the so-called *Himmelbett*, or four-poster bed, which was a radar-control set-up which could work in any weather. It used an improved long-range version of the *Wurzburg* radar, with stations set up along what became known as the Kammhuber line (after the architect of the system, *Luftwaffe* Major-General Kammhuber) guarding the approaches to Germany. Each station was assigned to a particular box of airspace. Within it the radar controller could select a bomber target and then give instructions to the night-fighter assigned to that same box. This would, it was hoped, bring it near enough to the bomber to pick it up on its own Lichtenstein airborne-interception radar and then shoot it down. By the end of 1941 the Germans had more than 300 night-fighters in units assigned to work with *Himmelbett*, and during 1942 more and more RAF bombers were shot down.

But British ground stations had picked up signals which hinted at the German airborne radars used by the night-fighters themselves. The only way to find out more was to send aircraft out as deliberate decoys. When a night-fighter found them and made an attack, sensitive equipment on the aircraft would monitor the signals sent out by the night-fighter's radar before, with luck, the decoy could take evasive action and try to escape the attack it had itself attracted. Night after night in the winter of 1942, Wellingtons of No 1474 Wireless Investigation Flight set off across the Channel, each one trying to draw the night-fighters into action. Seventeen times there was no reaction. Then, on the eighteenth sortie, the bait was snapped up. On 3 December, 1942, a Wellington crashed into the sea after being shot up by a *Luftwaffe* night-fighter, but not before details of the radar transmissions had been radioed back to the base in Britain.

At the same time, using new long-distance radars, the Germans were able to pick up echoes from British bombers as they took off and formed up over their home airfields. In addition, the *Horchdienst*, the *Luftwaffe* radio monitoring service, had become adept at reading the RAF plans from their figures of radio traffic during the day. Each day it was normal for operational bombers to carry out short test flights to test radio and other equipment; these routine transmissions were read by the Germans, and the pattern of the traffic analysed. An even density of

messages over morning and afternoon meant a raid that night was
unlikely. Very heavy traffic in the morning, with little in the afternoon,
meant a big raid was likely that night. There were also improvements
made to the *Himmelbett* network, increasing the depth and the effectiveness of the defences through which the bombers had to fly to reach
their targets.

The blow which was to begin the turning of the tables, and which was
to help more than anything else in crippling the morale of the night-
fighter pilots, was a by-product of that heroic Wellington flight. One
of the best fillips to bomber-crew morale would have been the presence
of some kind of fighter escort—but it was impossible at night to
differentiate between friend or foe. The answer was a system of devastating effectiveness called 'Serrate'. In its essentials, this was a simple
receiver which could pick up the German night-fighters' radar transmissions from a distance, and could then be used to lead the aircraft
which carried it straight to the enemy night-fighter. Carried in the
formidable Beaufighters of No 141 Squadron from the late spring of
1943, 'Serrate' began to change the whole balance in the night skies
over Germany. Time and again night-fighter crews would be engrossed
in the intensely difficult and demanding task of trailing a bomber
target to the point where they could see it clearly enough to open fire.
Suddenly, without warning, they would be raked by crippling fire
from a battery of four 20 mm cannon and six machine guns. With
about ten times the fire power of a bomber's defensive armament,
this was usually enough to destroy the night-fighter. Those who *did*
survive had frightening tales to tell their fellows, and to ponder for
themselves during their next sorties.

'Serrate' had another refinement—being a purely passive receiver
(in other words, not having to send out any signals of its own), it
could pick up German radar transmissions from behind the Beaufighter
as well as in front of it. So daring Beaufighter crews could station
themselves within the bomber stream, offering themselves as targets.
As the night-fighters picked them up and began to home on them, the
pilot could watch his adversary's progress on his instruments. Just as
the *Luftwaffe* pilot was about to begin his attack, the Beaufighter
would whip round in a tight turn and rake the German night-fighter
with fire.

This tactic produced an effect totally out of proportion to the losses
it inflicted. From the summer of 1943 onwards, every German night-
fighter pilot with a British bomber in his sights, wondered whether or
not he was about to become the victim. At the very moment when he

needed every ounce of his skill and courage, he was suddenly at his most vulnerable, and morale and efficiency received a crushing blow.

At the same time an even more effective navigation aid was coming into use with Bomber Command which had no need for ground stations at all. This was H2S, a small radar transmitter which showed the navigator a radar map of the country over which he was flying. Different types of echo from open ground, water, woodland or built-up areas, all helped to identify targets anywhere in Germany in any weather conditions.

The *Luftwaffe* defences received their most crippling blow on the night of 24 July, 1943. A heavy raid on Hamburg, involving 746 bombers, was seen by the German fighter controllers as involving more than eleven thousand aircraft! The cause of this illusion was a shower of tinfoil strips, called Window, each one thirty centimetres long by one and a half centimetres across, dropped in bundles of two thousand strips apiece. The effect of each shower of strips, the size of which had been carefully calculated in relation to the wavelength of the German radars, was to simulate the echo of an aircraft. Amid this enormous 'formation' of attacking bombers, ground controllers and night-fighter pilots alike found it was impossible to identify the real bombers which were seeding the night sky with the tinfoil bundles.

The whole efficient and formidable *Himmelbett* system was rendered totally impotent by one terrible mistake, a sin of omission by the people responsible for planning Germany's radar systems, which had placed this devastating weapon in Allied hands. The theory of Window demanded strips of metal of a size directly related to the wavelength and frequency of the radar which it was intended to neutralize. The main German early-warning radars, the fire-control radars for the flak batteries and even the night-fighters' airborne interception radar were close enough in frequency to suffer equally from the jamming. It was this fatal vulnerability which made the innocuous-looking strips of tinfoil fluttering down from each of the bombers such a deadly weapon.

For the time being there was no defence against the jamming. All the Germans could do was fall back on even more elaborate decoy targets, which could falsify even the evidence of the H2S displays. Back in March, 1943, an earlier attack on Hamburg had been foiled when the Pathfinder force had followed the course of the Elbe towards the city. Before they reached it, they had to cross a decoy site where the shape of the river banks was very similar to that in the city proper, and where army engineers had dammed a stream to produce a lake shaped very much like the Aussen Alster in the heart of Hamburg. One of the H2S-equipped bombers had a faulty set, and the blurred picture seen over the decoy site looked sufficiently like Hamburg for the bomb-aimer to drop his markers. And although other Pathfinders went on to drop their markers over Hamburg itself, ten miles to the eastward, the damage was done. Seventeen bombers actually attacked the city that night, although 344 were convinced they had. Most of them dropped their bombs on the wrong markers, convinced that the markers dropping over the city were German decoys—and the city survived virtually unscathed, although the little town of Wedel and several villages around the decoy site were obliterated.

This was scant comfort for the Germans, however. As the experience of the Pathfinder crews grew, along with their skill in using the new equipment, such a mistake was far less likely to occur again. Even the diversion of badly-needed anti-aircraft batteries to the decoy sites to put up convincing flak barrages, the construction of elaborate buildings and landmarks like rows of chimneys, and the development of rockets to produce flares looking like British markers in order to start the false trail, were no real use. More and more bombers were being equipped with radar, and those targets nearer to the coast were being marked by Mosquitoes using 'Oboe', the navigational error of which was too small for the pilots to be fooled by decoy sites anyway.

The Germans were still more confused by diversion attacks. Since Window gave a totally false estimate of the number of aircraft actually in a formation, small groups of Mosquitoes could simulate a radar picture of a much larger bombing force by dropping bundles of Window at many times the usual one-bundle-per-minute rate. This would lead the German controllers to send off all the night-fighters to chase the feint attack, leaving the real attack to be made free from opposition later on.

This worked well enough for a while; but before long the *Luftwaffe* signals branch developed a network of ground-based 'Naxos' trans-mitters which could pick up the pulses from the H2S radars of the bombers. Not only could bearings from several of these receivers be used to give a position fix for the bomber stream at any time, but it always told the Germans exactly which attack was a diversion and which the real thing. This was because the Mosquitoes used for diversions did not carry H2S, whereas the Pathfinder force leading a real attack always did.

Another modification was made during the summer of 1943 to help the night-fighters in their fight against the bombers. Since they could only find and attack their prey visually, it was equally easy for the bomber crew to spot their attacker and take evasive action. But the fitting of what was called *Schrage Musik*—two 20 mm cannon firing almost vertically upwards behind the night-fighter's cockpit—enabled the fighter to make its attack while flying below and just behind the bomber where it could not be seen and where the relative motion of the two aircraft was least, which made the aiming easy. For this reason the Germans were able to avoid using tracer rounds for the guns, so that bombers were shot down without even their closest neighbours in the bomber stream knowing the direction or the method of the attack. Bomber after bomber disappeared with no apparent reason, thanks to this savagely successful morale-breaking weapon.

But Bomber Command struck back in the next round of this psychological battle in the darkened skies, on the night of 26 October, 1943, when 569 bombers were routed to attack Kassel. This time the powerful Rugby radio transmitters were tuned to the German fighter-control wavelengths. Fluent German-speaking personnel began issuing false orders to the night-fighters, and the confusion was such that the genuine controllers nearly lost all control, and the bombers were able to reach their target with virtually no opposition. The night-fighters were able to rally in time to catch several bombers on the return journey, but the damage was done. Although the Germans went to great lengths to recruit women controllers and people with strong local accents difficult to imitate, the ever-present possibility of false orders was yet another worry for the sorely-tried night-fighter crews.

Yet just how dangerous a force the German night-fighters could be was shown by the disastrous Nuremburg raid on 30 March, 1944. Monitoring of H2S transmissions gave the lie to two feint Mosquito attacks on Kassel and Cologne, while allowing the ground defences to keep a close check on the progress of the main force on its long route to southern Germany. A total of 246 fighters was brought into the attack and, in all, the catastrophic total of 107 bombers were lost in a single operation, by far the majority from night-fighter actions. Yet diversions and deceptions could still tip the scales in Bomber Command's favour. The Essen raid of four days earlier had followed a route which hinted at a long-distance raid deep into Germany. Instead, they swung round without warning, bombed Essen and turned back for England with most of the night-fighters waiting for them further inland. Only nine bombers were lost, at least three from anti-aircraft fire, while the Germans lost sixteen of their night-fighters, mainly in flying accidents.

But the lion's share of responsibility for final victory in the night air battle must go to the RAF's special counter-measures unit, 100 Group, which was formed in November, 1943, to attack the German ground and air defences. By the time the bombers returned to the war against Germany late in 1944, the planes of 100 Group were on hand to produce a totally new kind of escort. Before any operation—and even on nights when no raids were planned, to avoid giving Bomber Command's intentions away, and to keep the Germans at full stretch—a force of just a few specially-equipped jamming aircraft would be sent up, each one briefed to orbit a selected spot outside German airspace. Between them, these patrolling jammers could blot out radar cover over a whole sector of Germany.

Meanwhile other 100 Group aircraft would mount several feint attacks to draw off the night-fighters. This time they would carry H2S sets to mislead the German listening stations, and drop Window of lengths calculated to blot out coverage on the SN2 sets of the patrolling night-fighters as well as the ground stations. As they reached their 'targets', the leading Mosquitoes would drop cascades of brilliant marker-flares. For the hapless citizens, convinced they were about to be subject to a full-scale raid of devastating intensity, this was a shattering experience, even when no bombs followed the flares. By then, all the defence and emergency forces were on full alert, and night-fighters would be on their way, while at that very moment a real raid was about to begin somewhere else altogether.

By this means the *Luftwaffe* night-fighters were persuaded to spend their time on endlessly frustrating wild-goose chases—like the night of 4 December, 1944, when Bomber Command mounted four full-scale raids, two on Karlsruhe and Heilbronn in southern Germany and two on Hamm and Hagen in the Ruhr. Almost 100 fighters of the *Luftwaffe* were sent by their controllers to counter a false raid on the central Ruhr area, with the result that less than two per cent of the bombers were lost that night.

Worse was to come for the wretched *Luftwaffe* aircrew; their new radars were now made the basis for an improved, longer-range version of 'Serrate', which was carried by fast, long-range Mosquito night-fighters of 100 Group. Another *Luftwaffe* improvement, an identification transponder which could send a coded signal when triggered off by an interrogation signal from ground control to identify the particular night-fighter, was to become an even more deadly Achilles heel. For the Mosquitoes were equipped with a transmitter called 'Perfectos', which could actually trigger off the night-fighters' transponders in the same way as the German ground stations could. This was another giveaway of the night-fighters' positions, and a positive identification of an enemy target.

By the final phase of the air war over Europe, the hunters of the night skies had themselves become the hunted. Whenever bombers were abroad, a squadron of 100 Group Mosquitoes would split up to patrol all the known night-fighter fields and the radio beacons where the *Luftwaffe* aircraft would orbit while awaiting instructions from ground control. Any night-fighter unfortunate enough to come within range was usually shot down, and the effects on the *Luftwaffe* were profound. They were in the greatest danger at the moments when they were most vulnerable—when landing, taking off or waiting for orders at a fixed reference point, and when they evolved techniques of their own for avoiding the Mosquitoes, like hedge-hopping just above ground level for hundreds of miles on end or—like the night-fighter ace, Hans Krause, who always approached his base by a power-dive from 10,000 feet followed by an immediate landing on a dimly-lit runway. By carrying out extreme manœuvres to throw the British aircraft off the scent, the pressure under which they were flying manifested itself in a steep rise in the accident rate. Since the fuel shortage had also begun to restrict night-fighter operations, only the best and most experienced crews were sent up in most cases, and these were the ones who mistook their bearings or altitude and crashed, or were shot down by their colleagues or by anti-aircraft guns because they dared not use their identification sets.

Within six months the long night battle was virtually over. The *Luftwaffe* night-fighter arm was finished as a fighting force, its morale shattered by the cruelly effective psychological battering it had received out of all proportion to the real losses inflicted by the long-range Beaufighters and Mosquitoes. By early 1945 the *Luftwaffe* was driven from the daylight skies by the enormous weight of Allied air superiority. Thanks to a brilliant combination of technical skill and tactical cunning, the German flyers were no safer or no more effective under cover of darkness. To all intents and purposes the air war was almost finished. Although there would still be occasions when devoted and courageous pilots would cheat all the odds to deliver desperate attacks, psychologically they already felt themselves to be defeated.

CHAPTER FIVE

THE RADIO WAR

IN THE Second World War radio naturally played a bigger part than ever before. Radio waves carried signals and communications between ships and aircraft, bases and units, headquarters and agents, on a scale never before experienced. So it seems particularly fitting that, from the very beginning of the conflict, the very ether itself should become another battle-ground, a theatre of war with its own weaponry, its own tactics and its own rules.

It began, as did so many other new ideas and techniques in warfare, with the Germans. Dr Josef Goebbels, that master-mind of propaganda, discovered at a very early stage that radio was a powerful weapon because it could be used to paint all kinds of pictures in the minds of its listeners, without its own origins ever being clearly established. The First World War had already shown the value of white propaganda — messages directed at the enemy in an attempt to wear down his morale and cause doubts as to his chances of victory; and although radio could intensify and improve the impact of this kind of psychological warfare, it had another potential as well. This was in the field of so-called black propaganda, a much more difficult, much more subtle and much more effective attack altogether.

Black propaganda was carefully disguised. Whereas straightforward white propaganda broadcasts from, say, Germany to England might concentrate on German advances and victories, new German weapons, British defeats and so on, its effect would be weakened because this was exactly what British listeners would expect an enemy station to say, whether it was true or not. On the other hand, if a British radio station were to produce normal pro-British broadcasts, the effect of any occasional hint of bad news would be greatly magnified, as its listeners would tend to accept it as the truth.

This is the essence of black propaganda. If the Germans set up a transmitter which sends out programmes which pass it off as a *British* transmitter, then most of the programmes will have to be exactly what its listeners would expect from a genuine British service. In other words, they will be anti-German, anti-Nazi, pro-British and pro-Allied. But selected items, carefully dropped in now and then, might hint at covered-up muddles in high places, catastrophic defeats not properly explained, and other well-proven morale-breakers. The balance is a very subtle one—too many gloomy items, or too much criticism, and the listeners may begin to realize all is not as it should be. Too few, and the impact may not be enough to make any difference.

The Germans began their psychological warfare offensive in the quiet of the Phoney War winter of 1939–40. In January and February, 1940, two secret transmitters began broadcasts against the enemy in the west. One was passed off as a secret station run by the French Communist Party, the other by anti-Government British based in Ireland. In this way there could be more acid criticisms of Government actions, and more attention given to Allied defeats, without giving the game away that the broadcasts were being put out by the Germans themselves.

But the secret transmitters really came into their own when the Western Front awoke from its eight-month sleep on the morning of 10 May, 1940. German forces swept across the borders into Holland and Belgium, Luxembourg and France. Paratroops seized key bridges and fortifications, dive-bombers paralysed resistance and shattered communications, to spread chaos and confusion over a wider area. These were exactly the conditions in which the secret transmitters came into their own. People were worried, suspicious and desperate for news. Official broadcasts, trying to stiffen morale, often did not give up-to-date information on enemy advances, and rumours were rife. Into this highly nervous situation the German secret transmitters began broadcasting news items about Fifth Columnists 'in the pay of the Germans'.

These Fifth Columnists were, according to the radio, to be found anywhere and everywhere, organizing acts of sabotage, treachery and murder to help the French defeat. Many of them, said the broadcasts, were in fact Germans who had crossed into France before the war in the guise of Jewish refugees—thereby raising hatred and suspicion against Jews who had genuinely fled from Hitler's persecution. The anonymous speakers cursed the incompetence and stupidity of the French Government and warned listeners that it was about to flee from Paris to refuge in the country.

As the Germans pushed deeper into France and the defending armies fell into headlong retreat, the confusion and panic grew worse. The German secret transmitter stepped up its 'Watch out for the Fifth Columnists' scare with deadly effect. The theme was taken up by the French newspapers and exacerbated an already tense and depressing situation. Morale began to crumble and people spent more time looking for traitors in their midst than guarding against the German advance from outside. Other broadcasts warned French listeners that the first action of the Germans after occupying a town was to confiscate all the cash they could find in the banks; the result was a run on the banks and a total economic breakdown in some areas.

Other campaigns mounted by Goebbels' clandestine radio station included attempts to stir up animosity against the British for their failure to give enough help in defending France against the German invasion. General Gamelin, the French Commander-in-Chief who had been replaced by Weygand, was hailed as a realist who had been sacked because he had proposed suing for peace in the face of the inevitable German victory. More curses against the French Government were backed up by hugely inflated figures of French losses, and a plea for a just peace now. The Government alone was responsible for incurring further suffering. At the same time the secret radio started broadcasting coded messages to non-existent anti-Government forces within France, the implication being that there was a wide body of French people dedicated to overthrowing the Government if it did not bring a speedy end to the fighting.

The crushing of France was executed with clinical efficiency. Harried from the air by Stuka dive bombers, cut-off and over-run by the phalanxes of Panzers, harried by indecision from above and the threat of Fifth Columnists in the rear, the army fought bravely but in vain. Prime Minister Reynaud, the chief target of the anti-Government broadcasts over Goebbels' secret transmitter, (there were now three, all supposedly located in France, but in fact broadcasting from German-held territory) resigned and Marshal Pétain took over. The Pétain régime began peace negotiations, but the secret transmitters kept up the pressure by emphasizing how bitter was the fighting which was still going on. Why didn't they end the whole business now that further resistance was useless? In the end the French did exactly that. And Hitler, with a mixture of bitterness and shrewd stage-management, insisted the surrender be signed in the same railway coach on the same spot as Marshal Foch had accepted the German surrender in 1918.

This left Britain as the main target for German psychological warfare. Goebbels himself ordered, on 22 June, 1940, that if the French accepted the conditions offered them by the Germans, the secret transmitter should close down, ending its last week's broadcasting on the hopeful note of how correctly the Germans seemed to be behaving, and that

there might well be better hopes of a new Franco-German understanding after all. Then, after a pause of a few days, it began a new
campaign against England, using a group of anti-Churchill British
prisoners of war recruited by William Joyce, best-known of the announcers in Goebbels' white propaganda team.

Hitler was well satisfied with his crushing victory over France in a
brilliant six-week war. Invading Britain promised to be a much tougher
proposition, and one far less dear to his heart than the challenge of
attacking Soviet Russia, an operation he dared not undertake until
Britain was less of a threat to his western frontiers. In the hope that
the British might be persuaded into talking peace terms, the radio
campaign was muted at first. The British forces, and the bravery of the
soldiers, sailors and airmen in the Dunkirk evacuation were all praised
to the skies, but Churchill was attacked as the man who was stupid
enough to keep a futile war going. The Germans already knew that
there was little hope of ceasefire with Churchill at the helm; but were
he to fall, a successor might be more amenable.

This caused Goebbels to step up the radio offensive against Churchill.
There were now no less than five different secret German transmitters
beamed against Britain. These included revolutionary socialist workers'
programmes, services aimed at Scottish and Welsh Nationalists, whom
the Germans always thought of as potential allies, and others aimed at
religious and pacifist groups. The campaigns varied slightly, but the
main theme was a common desire for peace. The pacifist broadcasts
concentrated on the advantages of Hitler's proposals and the worries
about what might happen if this opportunity were not taken, while the
'revolutionary workers' transmitter called for listeners to form action
committees to call for the replacement of Churchill as Prime Minister.
Other steps were held in reserve. Once the invasion went ahead,
thought Goebbels, the transmitters could start broadcasting fake
evacuation orders, which would produce total chaos in Britain and tie
down the defending forces.

As the peace initiative waned, Goebbels' secret radio transmitters
returned to the attack. Cleverly, he insisted that they prove their
'genuineness' to British listeners by scathing attacks on Germany and
the Nazis—but underneath this sugar-coating came the propaganda
pill. Discreditable stories about prominent leaders and politicians were
invented for the announcers to react to; eyewitness reports of the horrors
of war already suffered in places like Warsaw or Dunkirk dwelt on the
carnage involved in such a way as to imply that the British were not
being told the full truth by their own newspapers.

From the end of July new ways of undermining the confidence of the listeners were tried. A repeat of the French scare over the safety of savings was produced by advice to be prudent, and to build up as large a stock of foodstuffs as possible, (producing a run on the market), and the advisability of taking all one's savings from the banks and investing in gold or jewellery or other things of safe and lasting value. Instructional broadcasts, too, could be used to advantage. Experts were brought in to write a series of programmes on air-raid precautions, very much like the real ones put out by the authorities at the time, except that all the gory details of air-raid damage, dangers and casualties were so vividly described as to provide a real horror story. This was backed up ten days later by new programmes on defences against gas attacks, again stressing the horror of the gas threat, and subtly hinting that even the most careful precautions were not really that effective anyway.

But it was in the middle of August, 1940, with the Battle of Britain at its height as Hitler's invasion build-up went on growing, that Goebbels, switched the transmitters on to the next phase of the campaign. This was a repeat of the Fifth Columnist scare which had been so effective in France, but, because of the Channel, the story of infiltrators was less frightening. Instead the broadcasts warned of the dangers of German parachutists, pointing out that a hundred thousand British uniforms had been captured in the wreckage of Dunkirk. What was to prevent the Germans dressing up as British soldiers, parachuting into England by night and acting as spies, assassins or simple spreaders of confusion and alarm to aid the coming invasion?

This was psychological warfare at its best, because it reinforced what the British already believed. At the time it still seemed incredible that France, long regarded as Europe's strongest military power, had been laid low so quickly and apparently so easily. The lessons of the blitzkrieg had not been learned properly, and after the successes of Goebbels' Fifth Column scare, this came to be accepted as one of the great myths of World War Two. Stories of squads of parachutists, dressed as French soldiers, priests, civilians and even nuns, became part of the folklore of the times, and even before this latest campaign of Goebbels, the British had started to raise a civilian volunteer militia to cope with the menace they were convinced would soon be turned against them.

It may seem hard to suggest that the entire existence of the Home Guard, those brave and loyal volunteers who gave their spare time to training and drill without pay or reward, owed itself to nothing more than a propaganda threat, but the sad truth is that not only *were* no German Fifth Columnists landed by parachute, but the threat itself never existed outside the words of the secret transmitters. Apart from spies, who were invariably quickly caught because of the inadequacy of their training and German ignorance of conditions inside Britain, the only German parachutes seen over England were those belonging to shot-down aircrew baling out: apart, that is, from those which appeared in the north of England on the night of 14 August, after a broadcast which said that a massed drop of parachutists would end with the new arrivals being hidden by Fifth Columnists already in the country. Empty parachutes were dropped from German aircraft and many of them landed in cornfields, where the absence of tracks next morning showed that the parachutes had been dropped without human cargo.

Even so, the worries about parachutists went on growing in serious-ness. And undoubtedly a lot of time was wasted and confusion caused by road-blocks and security checks. Many perfectly respectable people minding their own business and doing their duty were suspected of being Fifth Columnists, and the Germans themselves helped the lie along by sending coded messages over their regular official white propa-ganda stations, as if intended for German agents within Britain. At the same time, as the bombing offensive went ahead, the myth that the secret transmitters were actually located within Britain was fostered by broadcast interviews with 'survivors' from some of the raids on British cities, who were able to dwell on the death and destruction caused by the German attacks, and imply that things were a great deal worse than official sources would admit.

Every measure which the British tried as a defence against air attacks was used by the Germans to frighten and worry people as much as possible. In the autumn of 1940 a new air-raid warning system was tried which kept the armaments factories working as long as possible when air-raids were expected, but not actually in progress. The secret transmitter aimed at the workers used this as the basis for an entirely fictitious news item about a factory being levelled, and all the workers being killed to save half an hour's production by not sounding the air-raid warning. This and other incidents were used to try and whip up workers' anger against the Government.

But by the beginning of 1941, with invasion of Britain replaced by an attack on Russia as Hitler's chief target, Goebbels was forced to admit to new troubles of his own. The minutes of his secret conferences for 7 January records him 'pointing to the great danger resulting from German troops listening in to British transmitters. There are indications that British broadcasts, which are now being put out daily from London for the German *Wehrmacht*, are being listened to on an increasing scale

by members of the *Wehrmacht* in the occupied territories. There is a
real danger of contamination by these enemy broadcasts.'

Goebbels went on to draw attention to the example of France, 'which
at the moment of decision had already been totally broken in morale by
German radio transmissions,' and gave instructions for an order to be
issued on behalf of the Führer, prohibiting members of the forces from
listening to enemy radios on pain of heavy penalties.

Imitation, they say, is the sincerest form of flattery, and for a past
master of pyschological warfare like Goebbels to speak in these terms
of British radio broadcasts shows just how successfully the British were
imitating the Germans. The British counter-attack in the radio war had
begun as early as the autumn of 1939 with propaganda and news broadcasts in German, but the impact of these transmissions really began to
strike home in the following summer. Goebbels' nearest opposite
number in England, Minister of Information Duff Cooper, recruited a
journalist named Sefton Delmer, pre-war Berlin correspondent for the
Daily Express, to the team of broadcasters.

Delmer was a great deal more than a skilled and experienced writer.
He had been at school in Germany during the First World War, spoke
the language better than many native-born Germans, and during his
years as a newspaperman had come to know many of the Nazi leaders
personally. So when Hitler made his conciliatory peace speech to the
Reichstag on 19 July, 1940, hoping for a favourable reaction from
listeners within Britain, Delmer was chosen to give the BBC's reply.
Two hours after Hitler's speech was recorded in England, he sat in
front of the microphone to deliver a stinging rejection in the rudest,
sharpest and most colloquial German. Hitler's magnanimous offer of
peace terms was hurled back, as Delmer put it, into his 'evil-smelling
teeth'. German officials, eagerly monitoring the British broadcasts for
signs of favourable reactions to the speech, were horrified and angry
at this brutal and cavalier reply. So were many of the British Government officials who felt that replying to Hitler was too important a matter
of policy to be left to the whims of mere broadcasters.

All the same Duff Cooper recognized the stuff of genius and two
results arose from Delmer's speech on opposite sides of the Channel.
The Germans put him on their black-list, due for arrest as soon as they
reached England, while Duff Cooper added him to the team of black-
propaganda broadcasters based in Bedfordshire and run by the Political
Warfare Executive.

At that time the stations which were beginning to cause Dr Goebbels
such concern – apart from the overt, or white, propaganda which was

beginning to win more and more listeners within Germany for the objectivity and accuracy of its news reports which provided the sugar for the psychological pill — were two separate black-propaganda stations. One was aimed at left-wing workers and was run by a group of emigré German Marxists. This was called the Radio of the European Revolution. The other was aimed at the right-wing opposition to the Nazis and was run by a former Weimar Republic politician now living in England.

Both these stations were superficially similar in their approach to Goebbels' secret transmitters; they were aimed at dissident-minded listeners within Germany; they tried to pass themselves off as genuine German stations broadcasting from within Germany, and they tried to stir up opposition to Hitler and his party. But when Delmer was put in charge of the right-wing station late in 1940, after the illness of its original director, he went several steps further — all of them, ironically, on lines already laid down by the Reichsminister for Propaganda and Enlightenment himself in his original offensive against England.

First of all, instead of appealing to his listeners to rise up against the Nazis, Delmer tried to make them believe that a large and highly professional resistance organization was already operating, and thriving, within Germany. But to make these resistance forces respectable, he followed another Goebbels' precept: the station was to be solidly patriotic from first to last. It would support the German Armed Forces, it would support even those Nazis who, like Hitler, were high enough in rank to enjoy the support of most Germans. Its real target would be the lower echelon of the party, who were bungling the direction of the war, letting down the soldiers and the people, and feathering their own nests at the expense of Germany.

But Delmer's most brilliant achievement was his attention to detail. The scheme would only work if its listeners accepted that the station was a genuine one, and to achieve this objective no detail which might help lend reality to the illusion was spared. The callsign was carefully chosen: GS1, or *Gustav Siegfried Eins* in German signallers' jargon, which sounded vaguely secret and vaguely military. As Delmer himself pointed out, in his book *Black Boomerang*, the initials could stand for General Staff 1 (*Generalstabs 1*), or Secret Transmitter 1 (*Geheimsender 1*) or many other more or less likely interpretations.

This of course was only half the battle. Every radio programme needs a presenter, and here too it was essential to find someone with the right kind of voice and the right kind of accent to make the listeners feel he was genuinely what he was said to be. Delmer was determined not to fall into the trap of using a stilted, gramatically correct but totally accentless speaker. Many of Goebbels' men gave themselves away by the carefulness of their English, and in any case the impact of what they said was reduced by the lifelessness of the material.

At last he found his man—a Berliner by the name of Paul Sanders, a one-time writer of detective stories who had left Germany in 1938 during the height of the persecution against the Jews, and who was now a corporal in the Pioneer Corps working on bomb disposal. Even this dangerous work had not blunted his urge to strike back at Hitler in the most effective way possible and he had volunteered for special service inside occupied Europe, which brought his name to the notice of the psychological warfare branch. His voice was almost perfect, deep and resonant, as Delmer describes it, with a slight Berlin drawl, just like the Prussian army officers who might be expected to broadcast on a station like this.

Corporal Sanders needed a name for his new image. Delmer decided on *Der Chef* (The Chief), a nickname he had often heard applied to Hitler before the war by his inner party entourage, and which again had the right military-secret combination to ring true for a conservative-loyalist resistance station. The signature tune for the new station was based on the answering phrase to that of the genuine *Deutschland-sender*, but instead of being played on the Potsdam garrison chapel carillon, it was hammered out on a cracked piano, as in a frontline billet.

All was ready for *Der Chef*'s first broadcast in May, 1941, when a ready-made subject for *Gustav Siegfried Eins*' first tirade against the Party bosses came down, quite literally, from the heavens. Rudolf Hess, Deputy Führer of Germany, flew to England in a Messerschmitt 110, and baled out over Scotland in an attempt to persuade Britain to agree peace terms with Hitler. Just twelve days later, with Germany still seething with news and speculation, GS1 went on the air for the first time—though the script was deliberately written to appear as if it were just another broadcast in a long-running series and that in fact *Der Chef*'s talk was in answer to questions from regular listeners.

Der Chef really let himself go. He hammered Hess as one of the 'cranks, megalomaniacs, string-pullers and parlour Bolsheviks who call themselves our leaders', a man who 'loses his head completely, packs himself a satchel of hormone pills and a white flag, and flies off to throw himself, and us, on the mercy of that flat-footed bastard of a drunken old Jew, Churchill'. Then he went on to attack Himmler as chief of security, claiming that to cover up his own incompetence—or worse—he was now arresting all manner of innocent and patriotic

Germans. By cunning deduction, GS1 gave a list of the most likely
arrestees, many of whom had actually been arrested on suspicion.
Der Chef ended his tirade with a vague warning of 'grave crisis' and
'dangerous developments' just around the corner.

This set the highly successful pattern for GS1. Attractive as the
proposition was, directly attacking Hitler and those around him was
out. Instead, GS1 went for the lower Party officials, who were known
to be doing a vital job in maintaining German morale and keeping the
mass of the people working solidly for the war effort, in spite of the
sacrifices which were being asked of them. *Der Chef* and his fellow
broadcasters were always solidly in favour of the top leadership, but by
convincing their listeners that the very people who were encouraging
them to give their all for Germany were contributing nothing themselves
but merely feathering their nests at the expense of the people, Delmer's
station achieved two objectives: it lowered morale and it gave every
loyal German an excuse, an incentive even, to do less than his duty.
Wherever scandals and rackets were exposed, full details were always
given so that anyone persuaded to follow suit would have all the
information he needed.

But the biggest problem facing any station like this in war-time was
to know exactly what was making the news within Nazi Germany – how
was rationing run, who were the officials responsible, what kind of
fiddles could be made to sound true? How, in short, could *Der Chef* be
made to sound convincing as a man in the know, and a man actually
inside Germany who spoke with authority? This was the stumbling
block which confounded many of Hitler's spies and Goebbels' broadcasters alike. The problem had to be solved if the broadcasts were to
have any influence at all.

Sometimes it was a matter of luck. Less than a month after *Der
Chef*'s dire warnings of momentous and terrible events to come, Hitler
attacked Soviet Russia. The timing was perfect. Clearly *Der Chef* must
have known all along, and must therefore have sources inside the
Government itself. But, for the tiny details which add together little by
little to build up a convincing picture, Delmer knew well that luck
would not be enough. He built up stories from careful studies of all the
German newspapers, from talking to refugees and others who had
come from Germany to work on the radio team. News announcements
in the German press, advertisements, trade directories, even the births,
marriages and deaths columns were scoured to build up a card-index
system of genuine people doing genuine jobs and living at genuine
addresses all over Germany. At the same time, hidden microphones in

the prisoner-of-war camps picked up all the conversations between newly-arrived prisoners and their colleagues who had been in England for some time. Although the speakers were often very careful to reveal nothing of any direct intelligence value, what they did say was just as valuable to the broadcasters: news from Germany, changes in the daily lives of ordinary Germans, new Government measures, new laws, new regulations, even new expressions and tit-bits of slang were all collected and used for the broadcasts. Mail from Germany to neutral countries which passed through the censors was combed for all kinds of information, and the whole vast mass of detail was filed and used to deadly effect.

The results were often right on target, like the great clothing-coupon scandal. Der Chef revealed that several women in Schleswig-Holstein were cashing all their clothing coupons at a number of department stores, because their husbands — Party officials who were, like their wives and the stores they patronized all real and all mentioned by name — had told them the demand for winter clothes for the army in Russia would soon be using up all Germany's supplies of cloth, so that the coupons would soon be valueless. Within weeks the scandal which Der Chef had invented had actually happened — a run in the clothing stores was being denounced, for the same reasons exactly, in a Kiel newspaper.

Another morale-cracking story was the Affair of the Blood Transfusions. This arose from a propaganda article in the German press on the Nazi medical organization, which praised certain units and personnel in the blood-transfusion teams by name. After careful thought, Der Chef put this information to good use in a totally concocted scandal, which said an old army comrade in the medical services (the doctor and his hospital were both named) had found out that blood supplies had been taken from Russian and Polish prisoner-of-war volunteers for giving to German wounded, and that much of the blood was infected with venereal disease and other unpleasant impurities. He cursed the doctors mentioned in the propaganda report as responsible for not having checked the blood first, and then, once the scandal had been discovered, of refusing to destroy the stocks of blood collected. As a morale-cracker, the story was deadly. Anyone who heard it would be worried at the thought of a blood transfusion for himself or any of the family from that day on.

But other stories were soon becoming accepted as true by the enemy themselves. For example, when a scrap of information reached Delmer's ears that Dino Alfieri, Mussolini's Ambassador to Germany, was shortly to be recalled to Rome for talks with the Italian Government, *Der Chef* was put on the air to demand Alfieri's recall. He quoted an entirely fictitious story about a named *Wehrmacht* officer coming home to his flat on leave from Russia to find Alfieri with his wife. Alfieri had been badly beaten up, but it was surely high time this Lothario was sent packing back to his own country. The result was that when Alfieri *was* recalled, the story was believed by all who heard it. Count Ciano, Mussolini's son-in-law, wrote that Il Duce was highly amused to hear the story about his ambassador being beaten up by his girl-friend's husband, even though the whole story was an Allied fabrication.

The same thing happened with another story, the testimony of a former servant of Robert Ley, boss of the Nazi labour organization, that the entire household was exempt from rationing because they qualified for Diplomat Rations, a scheme which really existed but which was meant to apply to foreign embassies and other Government departments with official entertaining to do. Because the term existed, and the alleged fiddle was an all-too-likely one, it was believed—so much so that Ley himself had to try and nail the lie in the issue of *Der Angriff* for 12 October, 1943, where he stated publicly that he and all his family lived on the normal rations and made no use of the Diplomat Rations Scheme. Yet in one of the prisoner-of-war cages, a newly arrived officer was telling his colleagues about how top Nazis were evading the rationing which applied to those actually doing the fighting and their families.

By now the Nazis themselves were becoming nervous about *Der Chef* and his uncommonly accurate sources of information. Although any direction-finder check ought to show that his transmitter was located in England, the legend grew that he was sending out his broadcasts from a variety of secret locations inside Germany. One persistent rumour placed him on a barge on the River Spree, travelling up and down between the interconnecting canals linking the Elbe and the Oder, to keep one jump ahead of the Nazi monitoring services. Even the Americans were fooled. They listened to *Der Chef* and came to think of him as hopeful evidence of new German Army resistance to Hitler, since he could surely go on operating only under Army protection?

Within its limits GS1 had done astonishingly well. But now the time had come to use these carefully developed techniques in a much more daring, but possibly much more effective, impersonation. Instead of a resistance radio, this would be nothing more than an ordinary German Forces' Radio station—carrying dance music, request programmes, news items, speeches by Nazi leaders and so on, in a perfectly genuine manner. But rather than the straightforward grumblings and criticisms of *Der Chef*, here the pyschological message would have to be dropped into the broadcasts much more subtly and much more gently. The majority of the items carried by the station would have to be entirely genuine, to provide the cover for the occasional lie. Very often the news itself would be correct, but a change in the comment, the conclusion or the emphasis would produce a very different impression in the listeners' minds. Other innocuous and genuine items would be insignificant in themselves, but by repetition, the message would eventually dawn on those who tuned in to the broadcasts regularly.

The new station was christened the *Deutsche Kurzwellensender Atlantik* — or German Short-Wave Station Atlantic, but the Germans themselves soon christened it the *Atlantiksender*. Originally aimed at the U-boat crews at the behest of the Admiralty, who were keen to widen the cracks in the submariners' morale about to be opened up by the new weapons coming into operation in the Atlantic battle, the station proved a deadly weapon. But it demanded much more expertise than the relatively simple requirements of *Der Chef* and *Gustav Siegfried Eins*. It needed the best possible dance music — German dance music at that — to make its potential audience *want* to tune in again and again to its programmes. Where could something as individual and ephemeral as popular music from an enemy country be found in wartime London?

Some of the music was genuinely German — hit records bought in neutral Sweden, and flown back in the converted bomb-bays of the BOAC Mosquito courier service between Stockholm and London. Other tunes were recorded in America, using German-speaking artists like Marlene Dietrich. But the U-boat sailors' own theme song, a parody of a popular dance tune, was specially recorded with a German vocalist, backed by the Royal Marines Band from Eastney Barracks, Portsmouth, in a session at the Albert Hall.

The announcers and presenters of the programmes had to sound genuine too, so the *Atlantiksender* team were fortunate in being able to recruit a varied selection of talent from the German Navy (and later the army and *Luftwaffe*) who had recently arrived at the prisoner-of-war camps and made themselves conspicuous for their anti-Nazi views. Not only did these men sound right, but they had the right background, the right technical knowledge, the right fund of slang and the right turn of phrase to sound genuine to their fellows in the same highly specialist branch of the service.

The background of the *Atlantiksender*'s appeal to its listeners, the sugar-coating on the psychological pill, was its request programmes. Producing genuine requests for genuine listeners involved another huge card-index file, with details gleaned from all the letters sent by U-boat prisoners-of-war to their families at home and the replies received from inside Germany. The U-boat fraternity was a close-knit one, and gradually a list of births, marriages, deaths and family gossip involving U-boat men still at sea was built up from the censored mail and from references in the German newspapers, and Agnes Bernelle, alias 'Vicky, the Sailors' Sweetheart', *Atlantiksender*'s lady disc-jockey, was able to send messages and dedications with the records referring to real birthdays, real anniversaries and real christenings, all of which persuaded the listeners that the station could only be genuine.

But the heart of *Atlantiksender* was the news broadcasts. These too had to be genuine, and here the team's trump card was a private direct line to Dr Goebbels' own offices for all the latest press releases from within Germany. This was a German *Hellschreiber* radio teleprinter which belonged to the London correspondent of Goebbels' news agency, the *Deutsche Nachrichtenburo*. In the rush to leave London as war broke out, he had left the machine behind and this meant that *Atlantik-sender* received all the latest German news at the same time as the genuine German stations: everything from Party leaders' speeches, war news, decorations and promotions to the latest results in the world of sport. And to add to the illusion, the station frequently broadcast speeches by prominent figures like Hitler or Goebbels in full. These were recorded from the genuine German broadcasts by the monitoring services and rebroadcast by the *Atlantiksender*.

In fact, all that distinguished *Atlantiksender* from its German opposite numbers, as far as its listeners were concerned, was the slightly different tone in which the news was presented. Where the normal forces' radio might mention that special 'air-raid bonus' issues of chocolate, then almost unobtainable in Germany, were being distributed in factory canteens to encourage workers to return to work as quickly as possible, *Atlantiksender* would drop in a detail of its own, explaining that the chocolate contained stimulants to help the overtired workers to greater efforts in spite of their fatigue. Where other German stations might mention the evacuation of children from the bombed cities to special camps in safer evacuee areas in the east, *Atlantiksender* would quote a speech congratulating the doctors at one of the camps who, despite the lack of drugs and antibiotics, had managed to reduce the death-rate from the recent diphtheria epidemic by sixty a week. Anyone who might want to break the law, to desert or to neglect his duty would never be encouraged to do so directly—but a long-running *Atlantik-sender* campaign tried to plant the idea that the German police were too old, too understrength and too overworked to cope with keeping law and order. A sports commentator would mention in the report of the defeat of a police football team that age had influenced the result. Rising crime figures were stressed, and the impossibility of coping with the situations caused by heavy air raids. The police themselves often did not know whether a man was missing because he had been killed or because he had deserted.

Other items came under the heading of 'listeners' advice columns', which also had their place in the psychological attack. A German High Command order of August, 1942, had been found among captured documents. It mentioned that servicemen whose families had been bombed out of their homes were entitled to compassionate leave. So *Atlantiksender* not only reminded listeners of their rights under this order, but the station also gave up-to-the-minute air-raid damage news as part of the service. This was made possible by a complex but brilliantly successful system which operated after every raid over Germany. Intelligence officers who helped to debrief the bomber crews on their return passed on details of what had been hit and where, how many aircraft had been involved and what type of bombs had been used. On the morning after the raid, RAF photo-reconnaissance Mosquitoes flew out to take pictures of the damage. These gave very detailed information, down to the streets and blocks of each city which had been destroyed. Copies of the prints were rushed by despatch rider to *Atlantiksender*'s researchers, who then compiled a list of streets from their collection of German street-maps covering every city and town in the country. So accurate were the reports of the streets and areas of towns destroyed and damaged in each night's raid that the listeners came to rely on them and to swallow the entirely fictitious, yet highly plausible, eye-witness reports of the horror which had ensued during the night's bombing. The German authorities, who knew all about *Atlantiksender*'s origins, were so convinced that this information could only have been supplied by agents inside Germany that they wasted much time and effort searching for spies who never existed.

At first *Atlantiksender* fooled many of the units it was intended to serve, even though Goebbels and his team realized immediately what was up. At least one German was up on a charge for piping *Atlantik-sender* to his comrades' billets for the fabulous dance music, not realizing he was committing the treasonable offence of listening in to an enemy station. And although Goebbels' frantic denunciations and warnings

7

began to have effect, so that gradually its vast and widespread audience came to realize it was indeed an enemy station, its effectiveness grew rather than diminished. For every time a request was played for a genuine event in a genuine U-boat sailor's family, the question arose as to how the enemy could know such intimate details, and what else did they know?

Delmer's men intensified this feeling among the submariners that they were being watched by asking the Admiralty's intelligence men in the field to send back any tit-bits of information they could about the German submarine service. Thus inter-flotilla football results, with all the scorers and details of where the team would be celebrating that night, were on the air over *Atlantiksender* within hours of the match finishing. When naval intelligence agents reported a particular U-boat as having sailed on patrol, Vicky played a special record to wish the boat *bon voyage*. To a U-boat, sailing under the strictest wireless silence and an absolute security blanket into the lonely Atlantic, it was a terrifying experience to hear that the enemy knew you had sailed, possibly even where you were at that very moment. How soon would the ping of the sonar pulses be succeeded by the blast of depth charges as the invisible attackers closed in for the kill?

So effective was this chilling ploy that the interrogators at the cages where newly arrived prisoners-of-war were questioned reported a marked change among U-boat men. Not only was morale much lower, but there was much less reluctance to answer questions, since they clearly felt that, after *Altantiksender*, the Allies must have access to all the top-secret information they wanted anyway.

Before long *Atlantiksender*'s success meant that it was taking over more and more of the psychological offensive from *Gustav Siegfried Eins*. Already operations of similar stations in the Mediterranean had introduced all kinds of details intended to prove to listeners, for morale purposes, that these stations really were operating from inside German-occupied Europe. In the case of one station supposed to be transmitting from Greece, this was periodically 'raided' by the 'Gestapo'. What really happened was that every so often pandemonium would break out in mid-programme, with shouts, screams, cries and the noise of overturning furniture, backed by sound effects of shots and machine-gun fire. Then the station would go off the air, only to reappear some days later, from a supposedly different and more secure location.

Now the time had come for *Der Chef* and *Gustav Siegfried Eins* to go the same way with more final results. At the end of October, 1943, the 'Gestapo' caught *Der Chef* as well, and his last broadcast ended in the clatter of machine-gun bullets from the sound effects department as a background to guttural shouts of Teutonic triumph. Paul Sanders himself was redeployed as a vital contributor to the expansion of the radio offensive in totally new theatres of war.

Radio programmes were transmitted to Hitler's allies in Bulgaria, Rumania and Hungary, but, of all his European allies, Italy offered the most tempting target. With the Allied victory in North Africa, followed by landings in Sicily and then on the Italian mainland itself, the country was ripe for defeat. This was a highly critical time, when careful negotiation or a shrewd psychological thrust could easily affect the whole course of the war. And a vital pawn in the Italian end-game was the cautious but splendidly equipped Italian Navy, by then confined firmly to its heavily defended harbours and anchorages in the north of Italy.

The danger which worried Allied naval chiefs, as it had after the fall of France, was that the Germans might seize the ships and use them to much better effect. So a station called *Radio Livorno* was set up — broadcasting from Allied territory, but pretending to be another resistance station like *Gustav Siegfried Eins*, only apparently based this time on one of the Italian warships lying in Livorno dockyard. The viewpoint of the unknown broadcaster, actually a Maltese officer in the British Army speaking perfect Italian, was violently anti-German. First of all he put his fellow-sailors on guard against any German attempts to seize their ships. Then, stage by stage, he assumed the role of the negotiator on behalf of the Italian Navy for their surrender, their freedom and their eventual swing to the Allies. Day by day his broadcasts were based on Admiralty instructions, and on 10 September, 1943, *Radio Livorno* radioed all patriotic Italian comrades throughout the fleet that the time had come. The fleet sailed for Malta in the teeth of a determined German attack which sank the battleship *Roma*, and there they surrendered to the Allies, exactly as ordered.

The second Italian radio operation was a straightforward but technically highly sophisticated counter-attack to a German radio operation. When Mussolini had been deposed and arrested, an attack squad of SS troopers had dropped by air on his mountain prison, released him and taken him back to German-occupied Italy to serve as a figurehead for what was left of his Fascist state. To assist his reacceptance by the Italian people as their rightful ruler, Goebbels set up a radio of the Italian Fascist Republic, which transmitted short-wave half-hour broadcasts over the Alps from Munich — each half-hour consisting of three sections, separated by music. This gave the Allied radio experts their chance. With a more powerful transmitter much closer to the area, it was possible for them to swamp the German transmissions altogether; but instead they aimed at a much more subtle tactic. When the end of the first section of the Fascist Radio programme gave way to the linking music, they readied themselves to take over. As soon as the music faded, the Allied transmitter was switched on, and two ex-Italian Radio announcers took over to send out a completely doctored news section in place of the smothered German transmissions. The monitoring service kept a careful check on the German transmissions in the meantime, and as soon as their middle section finished, the Allied announcers ended their programme, in time for the German music to lead into the genuine final item. So carefully was the change-over done that neither the listeners nor the Germans realized what was happening at first. But effects such as the sharp reaction to a scurrilous attack on the Vatican, or a run on the Italian currency caused by devaluation rumours, were blamed on Goebbels' own station, and the Germans were forced in the end to give way and cease operations on the Italian Radio front.

In the Balkans Allied subtlety went one better. The Bulgarian operation was mounted, using two German refugees, recruited in England, who spoke Bulgarian with a strong German accent. The programmes and the presenters were then made to sound like an extremely inept German attempt to fake a pro-Allied radio station — which insulted the Bulgarians and made the Germans themselves look ridiculous in an extremely effective psychological coup.

But the greatest achievement of the radio war was still to come. A few days after *Der Chef* died his heroic death at the hands of the Gestapo monitoring squads, another transmitter opened up from England against occupied Europe in the guise of a German Forces radio station. The *Atlantiksender* was a short-wave station, but the *Soldatensender Calais* (Soldiers' Radio Calais) used a far more powerful transmitter, broadcast on the medium wave band, and could be received all over Western Europe. This was the culmination of the radio war and it carried on all the psychological and morale-testing ploys of the earlier *Atlantiksender*, with some new and more subtle ones of its own.

When *Soldatensender Calais* first went on the air preparations for the D-Day landings were already well advanced, so an important aim of the new station was to imply to the German troops in the west that this was a backwater of the war, far from the battlegrounds where the real issues were being decided. This was done in two ways: by suggesting that only the best units were selected to defend Europe's eastern frontiers and that those left in the west were considered second-rate. Secondly, by implying that the crushing defeats in the east were not the fault of the German soldiers, but due to the enormous numerical advantages enjoyed by the Russians, backed up by the terrifying new super-weapons supplied by the Americans, like the phosphorous shells which could penetrate the strongest armour and incinerate anything, and anybody, inside. The obvious conclusion was not hard to draw: any display of outstanding ability, or even of average military efficiency, might result in a unit having to exchange the peace and quiet of life in the west for a mauling by the super-weapons on the Russian front.

Day by day news items reminded the German troops in France that more and more of their colleagues were being withdrawn for service on the Russian front at a time when the invasion was known to be imminent. Another news item might attribute the cause of a local defeat in Russia to Russian soldiers on the German side firing at their *Wehrmacht* 'colleagues' when the battle reached a critical stage, a report which made German units stationed next to Russian formations in France look at their neighbours in a new light.

Other reports stressed the sabotage raids carried out by the French Resistance dressed in captured *Waffen SS* uniforms, while others again advised the troops of alterations to leave trains caused by heavy Allied air raids on railway junctions in France and Germany. Instead of simply announcing the raids, as normal BBC broadcasts would have done, the *Soldatensender* chose the much more convincing line of advising its listeners where they could go for accommodation and entertainment while the damage to the junctions was put right.

The biggest campaign of all was the desertion campaign. Increases in the desertion figures were always being quoted, showing that more and more people were successfully getting away from the war. Mentions were made of the food parcels sent by prisoners in America, bought with their generous dollar earnings, and when cases of attempted desertion were quoted in condemnatory tones, the presenter was always careful to state exactly how it was done and where the men had made their mistakes—a direct encouragement to like-minded listeners to do better.

Once the invasion began *Soldatensender* came into its own. As the Germans realized what was happening, Delmer's men were able to use Goebbels' own press releases coming over the *Hellschreiber* link, embellished with all the details released to them by the Allied Supreme Command. From that moment on *Soldatensender* gave its German listeners accurate and factual situation reports which underlined the desperate situation the *Wehrmacht* was in; yet it became more and more trusted because its information, though unpleasant, was almost always true.

Throughout the post-invasion battle, the *Soldatensender* took the viewpoint of the front-line German soldier, betrayed by those who led him. Delmer's German workers broadcast moving tributes to German units cut off and overrun in Normandy. And when the V1 campaign opened, the *Soldatensender* cursed the flying bombs as a stupid waste of fuel incurred for the sake of a political and propaganda gesture – fuel which everyone in the *Wehrmacht* knew was desperately needed for air cover, or simply to save more men in the headlong retreat which began with Patton's breakout beyond Avranches.

But it was the bomb-plot of 20 July and the attempted assassination of Hitler which gave the *Soldatensender* its finest hour. As soon as the news of the coup came through on the *Hellschreiber*, the *Soldatensender* was busy encouraging army officers to support the rising to help secure peace before it was too late. Now at last, with the prospect of a Germany as divided in reality as they had always tried to make it appear in their bulletins, it was time to come out in the open and attack Hitler directly for the first time, although, as always, from a purely German Army standpoint. Hitler was the fool who knew nothing about soldiering, who relied on astrology, mysticism and intuition to take decisions which might cost the lives of millions of fighting men. And as Hitler's own supporters attacked the *Wehrmacht* for being implicated in the treachery of the plotters, the *Soldatensender* thundered out its reply, praising the ability, the war records, the selflessness and the good sense of the conspirators, in contrast to the greed and cowardice of those who were out for their blood. Every bulletin carried the suggestion that many more people were implicated in the plot than anyone had hitherto realized, making the apparent split in the German ranks wider and deeper hour by hour.

As Germany's situation worsened, such splits in the hitherto unbroken façade of Nazi rule became real enough. News filtered through to London of the delicate peace enquiries sponsored by Heinrich Himmler, Reichsführer of the SS, and Hitler's trusted replacement as Deputy Führer after the departure of Hess. *Soldatensender* started a campaign to blacken Himmler's name. All his speeches and articles were reported in such a way as to hint that he was courting popularity among the people. Other reports were angled or invented to hint at SS moves to seize vital stores and strongpoints as the war entered the Reich itself. Complaints were invented that Himmler was issuing orders direct to political officers in the front-line divisions without going through army channels. Field-Marshal Von Rundstedt was quoted as saying that the next step would be for Himmler to issue orders direct to units of the army as well as his own SS. Himmler's constant concern for the Führer's health, especially during his recovery from the effects of the 20 July bomb, was constantly mentioned. Leaflets and documents scattered by agents within Germany referred to Himmler's readiness to take over when Hitler became too ill to continue, and forged copies of the German soldiers' loyalty oath, with Himmler's name as Führer, were left in places which suggested they might have been mislaid.

Did Himmler ever realize what was being attempted against him? Possibly not, for in the end he proved the truth of *Soldatensender*'s campaign by trying to start independent peace negotiations, and Hitler sacked him before both men died by their own hands. Goebbels, another who was to commit suicide as Nazi Germany tottered, had no illusions about the threat which black radio presented. At the end of November, 1943, when the station had been on the air for less than a month, he confided to his diary that the *Soldatensender Calais* 'gives us plenty to worry about. The station does a very clever job of propaganda, and from what is put on the air, one can gather that the English know exactly what they have destroyed in Berlin and what they have not'.*

Even Himmler's own Security Service was alerted to the dangers of the German people falling for *Soldatensender*'s down-to-earth approach. A report, dated 16 March, 1944, from its Munich Headquarters pointed out that 'The chief effect of the station's news transmissions, which have been described as psychologically excellent, emerges from its practice of giving absolutely unexceptionable information, which has also been carried verbatim in the German News Service and mixing in with it a number of isolated, more or less tendentious items. This has caused large portions of the population to believe that *Soldatensender Calais* was a German station, perhaps one of the many *Soldatensender* started up in the occupied territories without anything about them being officially communicated to the population.' That the reports of the *Soldatensender Calais* often had a sharpness otherwise nowhere to be found in the German News Service was in some cases explained by the population on the following lines:

'After all they cannot present the soldier at the front with the same propaganda as they sell us at home. They have to be more honest with the soldiers at the front.'

'Since the New Year (1944),' the report went on, 'observers in

* *The Goebbels Diaries*, Hamish Hamilton, page 439.

Munich and the provinces point out with all urgency that the transmitter has caused the greatest unrest and confusion among the population by news concerning the situation at the fronts and at home and that
the population is showing ever-increasing trust in the station's news
service as its reports have shown themselves more or less correct. There
is general agreement that the majority of the opinions expressed among
the population concerning the situation at the front are derived from
the news of the *Sender Calais* which, in the words of a noted Munich
radio expert, belongs to the three most listened-to radio stations along
with Belgrade and the *Luftnotsender Laibach*.

'Politically responsible observers demand with increasing urgency
that action should be taken against this station with all means at our
disposal, above all that the population must be enlightened as to its
character as an enemy station. As this had not been done so far, the
population feels it has the right to listen to the station, on the one hand
because they cannot help listening to it on the Munich frequency, and
on the other hand because its effectiveness is not being interfered with
sufficiently.'

As it had not been possible to reduce listening to *Soldatensender
Calais* by confidential hints as to its origin, it was considered justified
that the station should be powerfully jammed.

But worried as they were about the station's effect on their morale,
the Germans were totally unable to shield their people from its influence.
For, as the report pointed out, the jamming succeeded in blotting out
the genuine *Reichsender Munich* service in the city area and in surrounding parts of Upper Bavaria. Yet with savage irony, not only was *Calais*
still clearly audible in those areas, but many of the listeners identified
the jammer worked by Goebbels' men against *Calais* as a British
jammer attacking the German *Soldatensender*!

As the war came home to Germany *Soldatensender Calais* assumed a
new and more direct influence on events. On Churchill's insistence, the
Allied commanders decided to cause as much panic among German
civilians as they could, driving long columns of refugees on to the roads
to clog the *Wehrmacht*'s defensive efforts, as those of the French had
been impeded in 1940. The Germans, however, were carrying out
evacuations only where they were absolutely necessary, and with their
own customary efficiency. So the *Soldatensender* was used to issue
counterfeit evacuation orders. The success of this deception depended
on the transmission engineers, who had now grown so adept in switching
frequencies and, with suitable scripts and sound effects, at cutting in on
genuine German transmissions so smoothly that hardly a break in

continuity could be detected, that there was no way in which the bewildered population could separate fact from fiction. Each night, as the RAF bomber stream headed for a particular area, German transmitters in that area would be cut off to avoid giving the bombers radio bearings which might help them fix their position. And because the *Soldatensender* team knew just where each night's raid was aimed, they knew which stations would go off the air, and even the approximate times at which they would be switched off. A team of announcers was built up who could imitate the voices, accents and rhythms of the regular *Deutschlandsender* speakers.

As the offensive gathered momentum, the careful German plans were swamped in total chaos. As soon as each transmitter went off the air, the *Soldatensender* team took over, breaking into the programme which had just been switched off, to issue evacuation instructions at regular intervals, just as the genuine stations did. Some instructions would tell the local people in a wide area to make for their nearest railway stations where special relief trains would be waiting with hot food and warm clothing. Other messages would order key Nazi party officials to leave their posts, close their offices and retreat to safer areas further from the front line, often hundreds of miles away, while other bulletins advised the population as a whole to make for several named areas in far-off central and southern Germany, where they could escape the bombing.

The result was that tens of thousands of people were out on the roads of Germany, making for fictitious rendezvous with trains and refugee columns which never existed, and local administration and resistance began to break down in the areas facing the Allied attack. Troops could not move along the jammed roads; panic was rife. The German authorities broadcast denials of the fake orders and repeated their own instructions, but once their transmitters had shut down before the approaching bombers, back came the *Soldatensender* transmitter to issue an identical denial, before repeating its own bogus messages, making the panic and confusion even worse.

This was the moment of defeat for the men of Goebbels' Propaganda Ministry who had first developed and used the weapon of black radio, who had refined it and evolved many of its classic ideas and techniques. Now, as in every other field, the men of the Third Reich found their own plans and tactics turned against them to crippling effect. With Hitler's empire sliding into oblivion, now was the moment for them to admit total and final defeat. Evacuation instructions were issued by cable instead; it was slower and often unreliable, but it was safe from

enemy interference. Yet by then the damage had been done; Germany
was alive with frightened people, marching they knew not where.
The Allies had won the Radio War as well as the military campaign
in Europe.

CHAPTER SIX

SUBVERSION AND SABOTAGE

AUTUMN, 1940, saw the Germans repulsed in their attempts to subjugate Britain from the air; but everywhere else they were masters of the Continent. In another nine months they would tighten their hold on the Mediterranean and the northernmost rim of Africa. But, apart from small-scale clashes in the desert and brave but hopeless rearguard actions in countries whose names appeared in steady succession on the *Wehrmacht* list of conquests, there was virtually unruffled calm in German Europe. Only the occasional pinprick here and there disturbed the euphoria of Hitler's Thousand-Year Reich. Yet, as the months went by, the stabs grew progressively sharper and more frequent. Not quite a pain, and not by any means a mortal illness, they were definitely adding up to a pressing irritation.

During this period psychological warfare became ever more important. For the time being it was the only kind of warfare which could be waged against Germany. But, however deadly the effects of each act of sabotage, each guerrilla operation, each assassination or even each moment of simple defiance, the material damage was still infinitesimal measured against the strength and the power of the Nazi machine. It was only the psychological effects which justified the efforts – the psychological boost to Britain and her undercover allies within Occupied Europe of striking back at the enemy in however limited a way, and the psychological damage done to the Germans by constant reminders that they were only holding down the hopes and the hatred of an entire continent by the strength of their fire-power. The slightest relaxation, the slightest inattention, and the street-corner or the café, the billet or the lane could become as lethal a place as a foxhole in the front line. Even a few commando raids, producing but a tiny fraction of the casualties resulting from traffic accidents or illness, and every *Wehrmacht* sentry from Bergen to Belgrade was having to peer anxiously over his shoulder, watching for any sign of movement in the shadows.

The beginnings of this nagging, undercover war were small and hesitant. For those under German rule the first real anger was kindled when the shock of defeat gave way to the realization that German occupation meant the complete exploitation of their country for the benefit of their conquerors. All national currencies were devalued against the mark, so that the German people became rich at the direct expense of their neighbours. Frenchmen and Belgians, Poles and Czechs were seized and put to work as forced labourers within the borders of the Reich. Food became scarce and harsh decrees quickly turned silent resentment and weariness into open fury.

At first retaliation took the form of snubs and jokes, designed to make the Germans look embarrassed or foolish. In Poland, where Nazi savagery was on an exceedingly short fuse, action of any sort was positively dangerous; yet all over Warsaw patriots fixed new street names in the course of a single night, rechristening the whole of the centre of the city after Polish heroes like Kosciuszko or foreign ones like Churchill. Elsewhere German propaganda posters were torn down or altered. One favourite, which carried the slogan *'Deutschland siegt an allen Fronten'* (Germany is winning on all fronts) was easily made to read *'Deutschland liegt an allen Fronten'*, (Germany lies helpless on all fronts). In French villages townhall clocks were stopped and tricolour flags hauled down to half-mast from the beginning of the Occupation. In Holland the entry of a German into a café or restaurant was the signal for everyone else to get up, pay their bills and leave. On the birthday of Prince Bernhard, ironically of German birth himself, but by now a leader of Dutch resistance, patriotic Hollanders wore white carnations in their buttonholes as the Prince himself used to do; the more cunning concealed razor blades beneath the blooms to lacerate the hand of anyone who tried to snatch the flowers away.

In Rumelange in Luxembourg a huge national flag was found flying from a factory chimney. When the Germans went up to cut it down they found that the last rungs of the ladder had been levered away, so that in the end they had to shoot it to pieces with prolonged machine-gun fire. At nearby Niederkron flocks of chickens were painted red, white and blue—and all over occupied Europe the initials RAF, standing for the only direct retaliation people could still see and identify with, sprang up on walls and buildings overnight. Frenchwomen began wearing red, white and blue striped dresses, tricolour ribbons in their hair and Cross of Lorraine brooches; their men sported the Free French emblem on watchchains and in buttonholes.

But the most effective of the early psychological symbols was the invention of a Belgian ex-minister, Victor de Lavelye—the sign of V for Victory. It had everything as a symbol: it was easily and immediately understood, its meaning was universal all over Occupied Europe, and it was made quickly by two strokes of a paintbrush or a wave of the fingers. Later C. E. Stevens, an Oxford Ancient History don on Delmer's staff, known to all as 'Tom Brown', pointed out the connection between the sign V in Morse Code (dot-dot-dot-dash) and the famous phrase in the opening movement of Beethoven's Fifth Symphony symbolizing the Fates knocking at the door. This was an ideal aural equivalent, which could be sounded on motor horns and engine whistles, factory hooters or even tapped out on a wall or a table. Used as the signature for BBC broadcasts to occupied Europe, the rhythm was tapped out on a single drum—a sinister sound, full of dread and portent which the Germans themselves admitted sent shivers down their spines.

BBC broadcasts were beamed into Belgium, Holland, Norway, Denmark and France. Apart from coded messages to specific resistance agents and units, they kept the ordinary people under German occupation informed as to what was happening in the war and the world outside. This was no propaganda network; the policy from the beginning was to tell the truth, however dampening and disappointing this might be at times. As a result the BBC broadcasts came to be looked upon as the only trustworthy voice to be heard. More than anything else, the radio helped European people feel that they were not forgotten, and convinced them they still had a vital part to play in the struggle against Hitler.

Valuable as the radio was, it was easily missed by much of the population in areas where tuning in to British broadcasts was a punishable offence. So the underground press set out to fill the gap with newssheets containing transcriptions of the radio transmissions. Taken down by undercover monitors, and either roneoed or turned out on a Letterpress machine belonging to a friendly printer, copies of sheets like the *BBC Bulletin* or the *London News* would appear mysteriously, to be passed from hand to hand. The Allies backed this effort up with their own newssheets which were dropped from aircraft—like the *RAF Post* or the *Voice of America*—but they were vastly outnumbered by the underground press proper. In Poland alone there were 1,400 newssheets produced at various times, more than 300 of which lasted out the whole five-year occupation. In France more than a thousand newssheets boasted a combined circulation of more than a million, and some of these productions were of the highest professional standards. After the Germans had taken over the Belgian newspaper *Le Soir* the underground press produced their own edition on 9 November, 1943 — a complete facsimile of the real thing, except that it was made up of

stories attacking the Germans and their collaborators. Distribution
was by squads of cyclists to kiosks all over the city.

As a weapon the press was limited in its effects, but it had two
major purposes: firstly it brought together whole populations who may
otherwise have had no contact with the outside world or with their
compatriots in the resistance or know what they were trying to achieve.
Secondly it drew attention to the crimes of the enemy—the injustices
of the occupation itself, and the sins of the collaborators. Above all,
it stressed over and over again that the Germans were going to be
defeated, however much it cost and however long it took.

All this helped to bolster the morale of those who were determined to
hamper the occupiers in every way possible. Direction signs were torn
down, telephone lines cut and petrol stocks contaminated, everything
possible was done to cause the maximum disruption and delay. Key
workmen arriving for duty a few minutes late each day would not
attract reprisals, yet a whole factory might be held up for that time
each morning, wasting millions of man-hours a month. Machines broke
down through incomplete servicing or grit in the bearings; trains
arrived late through real or imaginary defects. The railway rule book
became a powerful weapon in its own right. Armed with one, a local
official could use the most trivial incident to halt or divert a train, or
order its load to be transferred to a different route or reloaded on
different wagons.

Where over-zealousness left off deliberate inefficiency took over.
Orders were mislaid; wagon-loads of vital components were attached to
the wrong trains and sent to the wrong destinations. One load of vital
bridge-repairing components for north-eastern France ended up in
Guernsey, because of sabotaged instructions, where it stayed for almost
a year before the error was traced and a special convoy sent to carry it
to where it was needed.

Letters for the troops, especially for the pro-Nazi volunteers in
Waffen SS units or in the labour battalions, were lost by postal workers
—a heavy blow to morale. Replies were scanned for any information
useful to the Allies. Telephone cables were cut and reconnected by
repair gangs to allow for regular eavesdropping on *Wehrmacht* signals.
Doctors helped to cut the labour force by classifying fit men as sick or
disabled.

The advantage of this kind of warfare was its psychological impact
on the Germans. An air raid, however severe, was a familiar hazard
and one they could come to terms with, especially because there was
usually some warning. Any Frenchman, Dane or Dutchman could be

a saboteur, and the most innocent object an instrument of sudden death.

This advantage was exploited to the full in Special Operations Executive's laboratories, which produced a brilliant armoury of bombs and booby traps to confuse and frighten the enemy. There were sausages of plastic explosive which could be stuck to the panelling of aircraft, and magnetic bombs for use against tanks. There were imitation animal droppings — plastic explosive carefully faked to look like animal dung, using advice from the experts at the Natural History Museum to make it as lifelike as possible. A resistance man could, in countries where horses were used for transport (and this applied to most of occupied Europe), always add one or two of the fake droppings to every pile he found on roads used by the Germans. Any German vehicle driving over the harmless-looking pile would set off the explosive. Similar fake mule droppings were made for use in southern Europe, along with camel droppings for North Africa and even elephant droppings for the Far East.

Other sabotage devices were designed for resistance agents to use on machinery in their care. Carefully disguised explosive nuts and bolts could be fitted to machine tools, railway engines or vehicles, to blow them to pieces at the moment which would cause the maximum disruption. But perhaps the deadliest weapons from the psychological point of view were the bombs designed to be set off by the Germans themselves. These included genuine-looking milk bottles which exploded when the top was opened, loaves of bread which blew up when cut or broken, cigarettes which exploded on lighting, and pieces of coal or logs of wood which went off as soon as they were put on a fire. All these things could be left as part of normal deliveries, and the saboteurs had every chance of being miles away when the explosion occurred. Yet for the Germans, every bottle of milk, every loaf of bread and every piece of fuel would become suspect, causing endless trouble and deadly fear in even the most routine actions.

The first German sentry to be shot by the resistance was killed in the Somme region as early as 28 June, 1940. But German reprisals were savage, quick and indiscriminate, and the deaths of many innocent people would, as the Germans intended, produce a backlash of public opinion against the resistance, as well as the immediate intensification of hatred against the Germans themselves. So from a psychological point of view as well as an operational one, SOE believed it was better to wound the enemy rather than kill him. Deaths could always be hushed up to prevent damage to morale, but a wounded soldier could not be hidden, tied up badly-needed hospital resources and would tell everyone he knew about what had happened. When Germans *were* killed, agents were briefed to make the killing look as if it had been done by another German, to spread as much mutual suspicion among the troops as possible.

This bitter warfare was backed up by an increasingly effective barrage of ideas and aids developed in England, aimed at destroying morale by creating suspicion in enemy servicemen that those in authority above them were incompetent and corrupt, demanding that the fighting men make all the sacrifices while the heirarchy were free to make maximum profits from their efforts. At the same time the thought was planted that among a man's own comrades were would-be deserters or enemy agents. On the other hand desertion was encouraged as the simple and sensible thing to do for men faced with horrible, inevitable and utterly pointless death before Germany's final defeat.

The first of these campaigns made use of one of the Germans' own morale-boosters—a regular newssheet called *Mitteilungen fur die Truppe*, which was issued to officers so that the information it contained could be passed on to their men. A highly accurate copy, correct in layout, typeface, printing and writing style, was turned out in England and distributed to agents in France. It set out to warn officers about the huge increase in desertions to Sweden and Switzerland and it instructed them to order their men to help the completely overburdened field police in coping with the hunt for these dangerous men, to stop them giving a false impression abroad of the *Wehrmacht*'s fighting spirit. The leaflets were carefully dropped where a German officer might have mislaid them, where they could be picked up and read by the troops themselves, thinking the information was genuine but not intended directly for their eyes.

Convincing the Germans that desertion was rife was part of the battle. How to convince them it was also easy? This was the role of another counterfeit document, called *Krankheit rettet* or 'Sickness cures'. This was printed in a number of guises, as a German Navy handbook on fitness training, a railway timetable, a hymn book, and so on. In each case the first few pages were entirely genuine, but the meat of the book lay in a section containing careful instruction on how to feign the symptoms of an illness well enough to avoid unwelcome duty or a dangerous assignment. It told the reader step by step exactly what he should feel and how to lead the doctor on to make his own diagnosis, which would convince him the patient was telling the truth.

This was a deadly weapon from two points of view. In the hands of a man already wanting to avoid duty each copy was the means of reducing German strength by one or more men. But in the hands of the authorities its impact was far worse. If this kind of information was known to be freely available all over Germany, then everyone reporting sick, however genuine his symptoms, would be suspect. Honest and conscientious men were reprimanded and sent back to duty, causing bitter resentment, a sharp drop in efficiency, and a rise in infections among their comrades.

Later, these medical weapons grew more ambitious and sophisticated. Packets were dropped in German-occupied areas which gave the finder everything he needed to counterfeit a permanently disabling disease like tuberculosis, preparations which could change his temperature and blood pressure, fog X-ray plates in the right way, and produce an entirely convincing diagnosis.

The next step was to convince individual Germans that successful desertion carried all kinds of very real benefits with it. For example, many German families of men posted missing continued to write to their menfolk in the hope that they might still be alive and well in a prison camp in England. These letters were collected and replies were sent, posted by agents inside Europe, telling the families that their missing men were indeed alive and well, but in neutral countries, and not to attempt to contact them for the time being. As the grip of the Allied economic blockade tightened around Germany, the story was spread that prisoners in Allied hands were earning good wages and living off the very best. To back this up, generous food parcels were sent to the families of missing men and genuine PoWs alike through the neutral countries.

At the same time no opportunity was lost to discredit the Germans' own leaders and weaken the bonds of loyalty which kept many otherwise weary and dispirited men from actively thinking of desertion. One brutal but effective campaign centred on the German practice of sending signals from military hospitals to the local Party Officer of families of men who died, asking him to break the news direct. These messages were intercepted by British monitoring stations, and once again the forgers set to work. This time they produced letters on the correct hospital notepaper, which appeared to come from a fellow inmate of the dead soldier, sympathizing with the family and hoping that the diamond brooch or the gold crucifix he had bought as a gift had reached them safely through the Party officials.

The implication was that the Party bosses were robbing the families of dead soldiers. No one would dare spread this kind of accusation

8

abroad, but gossip within the family and among friends would be bound to spread like wildfire. Other letters carried the same theme to even more cruelly effective lengths. The letter from the comrade in the same hospital would imply that the soldier might have made at least a partial recovery, after a long convalescence, but he had been given a fatal injection as the bed had been so urgently needed for less serious cases who could still be put back into action to go on fighting for the Party.

Another letter caused a furore within Germany. When the young fighter ace, Werner Molders, was shot down in error by *Luftwaffe* anti-aircraft fire while coming in to land at an airfield near Breslau, several factors were left unexplained. Among them were Molders' increasing disenchantment with the régime and the war, and rumours that his attitude had led to the security services opening an investigation. A letter was written in England, as if from Molders, to his friend, the Roman Catholic Bishop of Stettin. In it he spoke most movingly of how many of his comrades had died in vain and how many had sought help and comfort in religion. He referred to the Nazis as the Godless Ones, and he asked for absolution, as he felt his own days might be numbered. The letter was printed on immaculately-forged *Luftwaffe* signal sheets, together with an introduction supposedly written by another *Luftwaffe* officer, enraged at what he described as Molders' murder on Nazi orders.

This was psychological dynamite. All over Germany the letter became a vital subject of discussion. It provided anti-Nazis and enemies of the régime with deadly ammunition. Furiously, the Propaganda Ministry tried to denounce it as a fake. Everyone from Molders' mother to Dr Goebbels himself was wheeled in to expose the lie, but so skilfully did the letter fit the facts that doubts remained for ever afterwards.

SOE agents brought back up-to-date copies of German travel permits and ration cards and these were printed by the thousand to be dropped all over Germany. Desperately the Germans responded with new issues, but these too were counterfeited and dropped. Goebbels himself countered with a different ruse; he had some very clumsy forgeries specially made and these were shown all over Germany as examples of how easily detectable the forged cards were. Those who were caught using forged ration cards were given exemplary sentences.

Other kinds of sabotage were more direct. Three-inch incendiary tubes called 'Braddocks' were dropped over Germany by American bombers. With them went leaflets encouraging foreign workers to use them to sabotage the factories in which they were employed. The idea was not so much to use the devices, which were of doubtful efficiency anyway, but to cast suspicion on the foreign workers. Huge searches were laid on to find the incendiary devices, and they were often blamed for fires actually caused by incendiaries dropped by bombing aircraft.

Posters were put up by agents in U-boat bases, telling the increasingly harassed U-boat crews how to earn themselves another six weeks in port by a bit of simple sabotage. Again the idea was not so much to persuade them to do this as to alert the authorities, so that the increased suspicion and extra security checks would cause another drop in morale. Other ideas were less subtle. Agents in Norway were given phials of liquid which produced a terrible smell when spilled on clothing, rendering them absolutely unwearable afterwards!

Many of these campaigns must have had an appreciable but unmeasurable effect on German morale. In Poland, most harshly treated of all the subject nations, the sabotage and subversion programme reached epic proportions. In the five years of German rule, 1,300 trains were wrecked, more than 7,000 locomotives destroyed and 24,000 wagons blown to pieces. From 1943 onwards all German trains to and from the Russian front had to be heavily escorted.

Even in quieter occupations, like that of Denmark and Norway, the resistance had an electrifying effect on German strategy. In the last few weeks of the war, when the Fatherland's fate rested with the last half-vanished battalions of Hitler's ghost armies, eighteen strong, fit and fully equipped divisions were trapped in enforced idleness, purely to counter the threat of the Scandinavian resistance movements.

The strong psychological backup had its effects too. One of the greatest compliments to its potential was to be dropped across the Allied lines as their armies closed in on Germany, in a last forlorn hope that propaganda could still achieve alone what dwindling material and manpower could not. Leaflets were picked up all over the area occupied by the Allied armies telling the soldiers how they could avoid dangerous frontline duty by skilfully feigned symptoms. Prepared on Goebbels' orders, using valuable plant and material, at a time when neither could be spared, they were direct translations of the 'Sickness cures' leaflets scattered throughout the Reich.

CHAPTER SEVEN

PREPARATIONS FOR INVASION

SOME MONTHS after Alamein, with the Afrika Korps falling back to the Mareth Line, the Allies decided to try and catch Rommel in an enormous trap by landing another army behind him. The British First Army, backed up by American forces which included the elite First Armoured Division, was to land in Morocco, an undertaking which demanded ship and convoy movements on a scale which could never be kept hidden from the Germans.

So the Psychological Warfare Branch transmitters began the first strategic deception campaign. Transmitters in England sent warnings to the population along the northern coasts of France, ordering them to move inland for their own safety, while the Vatican was informed through the Political Warfare Executive of likely operations in the areas of the Pas de Calais, Flanders and Normandy, so that precautions could be taken to safeguard religious treasures. The RAF built up the scale of its attacks all along the French coast, and Commando raids helped to keep the defenders on full-scale alert.

So far none of these deceptions really made an operation like the Dieppe raid impossible. Deliberately, they pointed to far too many areas for the Germans to fortify or reinforce all the threatened zones, and still allowed for local surprise at the actual point of attack. But by August, 1942, Hitler was so convinced that the Allies would land in northern France that he was actually withdrawing troops from the Russian front.

Unfortunately this theory fell down once the Germans managed to deduce where the attack would come. The *Abwehr* had already penetrated part of the French Resistance network in the Dieppe area and they knew the text of the message which would be broadcast by the BBC on the evening before a landing to alert the local Resistance. They also knew that a French contact in the German coastal fortification organization was supplying the Resistance with information on German defences — but they let the contact continue, since the areas for which the British requested information would give them an accurate picture of Allied intentions. Sure enough, in came requests for information on the defences of the town and harbour of Dieppe, and the Germans knew where (and, when the radio message was broadcast, when) the attack would come.

So they embarked on a deception scheme of their own to conceal the fact that they knew what the Allies were planning, and to make Dieppe a more tempting target than it really was. Carefully planted information was fed to their Resistance contacts, hinting that the garrison consisted of a single battalion of untrained replacements, attached to the badly mauled 110th Division, resting and refitting after service in Russia. In fact there were three battalions of the first-line 302nd Division in position, with three more battalions in reserve, with supporting artillery and a full Panzer division within call.

When the Dieppe raid was launched the Germans were ready. Alerted both by the BBC code message and the sudden drop in wireless traffic which often preceded an Allied attack, they were also able to read some of the naval signals in a cipher which the German cryptographic service had broken. Huge minefields had been laid in secret, guns had been hidden in caves, safe from aerial reconnaissance, and tank crews readied for immediate support.

The result was that the Allies' own deception plan rebounded against them. When the troops went ashore in the dawn of 19 August, 1942, they expected to meet only a half-hearted defence. Instead, they crashed head-on into one of the most alert and experienced units of the German Army. The landing was stopped on the beaches and more than half of the 6,000 men involved were killed, wounded or captured; German losses were less than 600 killed or wounded.

Yet in a sense Dieppe was to become a tragedy for the Germans too. So crushing was their victory that they were to over-estimate their chances of beating off any further Allied landings. Their own propaganda, which inflated what had been a limited-scale raid into a serious attempt at invasion, caught them in its own trap, leading them to ignore or to underestimate the much greater danger two years later.

The Allies, on the other hand, were to learn priceless lessons from their all too obvious mistakes. Meticulous preparation, painstaking planning, bulletproof security and, above all, a Chinese puzzle of deception within deception in seemingly endless succession would ensure the *Wehrmacht* never enjoyed such an advantage again.

Any deception concerning the landing in North Africa could only have a limited usefulness, however. In the early stages it shielded leakage of information about landing craft or the assembly of troopships and

their escorts. But sooner or later the course taken by the invasion
fleet would make it obvious that its destination was the Mediterranean
and not the Channel. Since this would still leave the Germans ample
time to prepare their defences, other suggestions were carefully leaked
that the Allies might first make a feint attack in North Africa, purely
to concentrate Axis attention there so that the main blow would catch
them unawares. The idea of the Allied plans, said these messages, was
to make the Germans commit badly needed reserves to North Africa,
where they would be unable to affect the fighting in France.

It worked beautifully—but only for a time. When the first Allied
troops came ashore in Morocco Vichy resistance crumbled. The
Germans did little at first, convinced that they had already been forewarned. They had been told of deception landings in North Africa
as a first step. Now it seemed even more certain that the main Allied
blow was on its way. So precautions were stepped up along the northern
coast of France.

But as the Allies went on to seize all of Morocco and then Algeria, it
became impossible to keep the deception going. More and more men
came ashore, with more and more ships arriving to supply them, until
slowly the truth began to dawn on the Germans. If this *was* a deception,
it was big enough to be dangerous in its own right. Desperately the
Allied command struggled to assemble a force strong enough yet fast
enough to race eastwards along the coast road to take Tunis and
Bizerta, Rommel's only lifelines since the Eighth Army had retaken
Tripoli.

All that could be spared from the landing forces was a scratch force
of British infantry in lorries, backed up by a solitary regiment of
obsolete cruiser tanks, Valentines and Crusaders, which were woefully
outgunned by the later types of German Mark IIIs and Mark IVs.
Close behind came a combat group from the American First Armoured
Division with lightly armed but fast and manœuvrable Stuarts. Full
of hope, this motley force set off eastwards. If they could reach the
ports before the Germans arrived there was a reasonable chance they
could hold them until reinforcements came. Cut off from their suppplies,
the whole German Army in North Africa would have to surrender
within weeks.

But geography was against them. From western Algeria to Tunis
was more than five hundred miles over appalling roads. Supply trains,
airfields and air support were left far behind as they ploughed onwards,
across the frontier into Tunisia and down the Medjerda valley. Yet it
was already too late. The Germans had finally awakened to what was

happening and, when it came, their reaction was not, as the Allied planners had fervently hoped, too little and too late. Late it may have been, but effective it most definitely was. With Sicily only half the distance from Tunis it is from Tripoli, shiploads of tanks were already unloading at the very quays which the British had hoped to seize, and squadrons of troop transports were shuttling backwards and forwards from airfields only hours away. Reinforcements, in short, were pouring into North Africa on a scale which, had they arrived six months sooner, would have taken Rommel across the Suez Canal. Led by the 10th Panzer Division, backed up by a single regiment of the new Panzer VI Tiger tanks, armed with the 88mm gun which had already proved such an effective weapon in desert battles, this new Fifth Panzer Army was streaming ashore as fast as German efficiency and energy could contrive.

The two armies' advance guards clashed at Djedeida, only a few miles to the west of Tunis, but it may as well have been fifty or a hundred miles away. Against the skilled Germans, with their immensely superior forces, the advance was stopped dead, although the Americans, facing a gap as yet unplugged by the Panzers, managed to shoot up the German dive-bombers parked on Djedeida airfield before being pushed back.

The loss of the race for Tunis cost the Allies another six months' bitter fighting. And if the Germans could hold on so long with what were still meagre forces compared with those lying in wait on the European mainland, what were the prospects for invasion now?

Yet at least the deception plan had worked, if only to a limited extent. If the Germans had realized sooner what the Allies were up to, they might have stopped them much further back, winning themselves a longer respite still. Conversely, had the Allies been able to keep the Germans guessing a little while longer, Tunis might have fallen six months sooner than it did, with incalculable consequences for the course of the war.

This made 1943 a crucial year in the Mediterranean, for outguessing and outwitting the enemy became supremely important to Germans and Allies alike. The British and the Americans had the initiative, in that only they knew when and where they would attack, but success could only be theirs if they could keep their intentions secret from the enemy. The Germans, on the other hand, would be fighting on their own ground, in prepared defences, with better communications and better supplies, at least in the vital initial phase. Furthermore, they knew well that any landing successfully repulsed would ensure that any further attempts at invasion would be postponed for a long time.

For the Allies the next target was to be Sicily, within reasonably easy reach of North Africa, and an obvious stepping stone to landings on the Italian mainland. But this was as obvious to the Germans as to anyone else. What possibility was there of convincing them the blow might fall elsewhere?

Greece was the only real alternative, since after the North African landings the German armies occupying northern France had taken over the southern part of the country too. Yet too much talk of Greece and the Germans might conclude that this was deliberately designed to distract their attention from Sicily. But drop hints which were too light, too subtle or too complex, and they might be ignored altogether.

What was needed was a piece of information so obviously not a deliberate plant, and so definite in its import that the Germans could not possibly ignore or disbelieve it. Once this had been done, if it could be done, then any other references to Greece would be picked up and added to the picture. So it was that Operation Mincemeat began to develop in the minds of British Naval Intelligence. The theory was that only a really secret message would carry enough weight with the Germans, but how could they be given the information? Simply letting them intercept a radio message was far too obvious. All the most secret messages were carried by courier to ensure strict security. But what if a courier were to be shot down and his body were to be washed up on an Axis shore, carrying a parcel of documents pointing towards a landing in Greece? Provided this could be done convincingly enough to make the Germans believe the body was genuine, then they would have to accept the information as genuine too. On the other hand, if they saw through the deception it was tantamount to telling them formally that landings would soon take place in Sicily.

First priority was the right kind of body. After a prolonged search, the corpse of a young man who had died from pneumonia was found, in a condition which would be similar to that of a drowning victim. A cover story was built up; he was dressed in the uniform of a major in the Royal Marines and given a series of carefully drafted letters from commanders like Lord Louis Mountbatten and the Chief of the Imperial General Staff to officers like General Alexander holding commands in the Mediterranean. All signatures and references in the letters were genuine, except for the mentions of landings in Greece rather than Sicily.

But simply telling Alexander that the landings were to be in Greece would be too much for the Germans to swallow. Would not a top commander like Alex know that already? Instead, reference was made to an agreement reached over extra reinforcements for the landings at Kalamata and near Cape Araxos, as if these were part of a detailed plan already known to Alexander. Then details were given of the Allied deception plan to make the Germans think that *Sicily* was the target — including amphibious training exercises to be held off the Tunisian coast and heavy air attacks on the Sicilians airfields, all of which would have to be carried out anyway as part of the preparations for the real invasion and which would help reassure the Germans of the genuineness of the information.

A final master stroke was referring to the Greek invasion plan as Operation Husky. This was the real code word for the Sicilian invasion, and any references to it which German intelligence might pick up would tend to lead them still further in the direction of an attack on Greece. At the same time, when their reconnaissance confirmed that exercises *were* being held off Tunisia, and when their airfields in Sicily *were* attacked by the Allied air forces, these measures too would seem to be part of the same plan.

On 30 April, 1943, the body was lowered into the sea from the deck of a British submarine, at a spot off the Spanish coast where currents would carry it ashore near the port of Huelva, where the German diplomatic representative was known to be specially alert. A fortnight later, following strong British diplomatic protests, the body was returned, together with all the documents, still in their sealed envelopes. Careful laboratory examination showed that the letters had been steamed open, read and resealed.

In time the full effects of these vital letters became known: extra German troops were rushed to Greece and Crete, including the crack 1st SS Panzer Division under General Sepp Dietrich, together with Sicilian-based naval units which might have helped to hold off the real attack. An SS Brigade was sent to Sardinia and a Panzer Divison from southern France to Greece, with two others on the Russian front being put on readiness to move. Rommel set up his advance headquarters in Greece on the very night that the Allied invasion forces were approaching the coast of Sicily.

When the genuine landings *did* take place, just ten weeks after the body had been washed ashore, only two German divisions were there to meet it, alongside an Italian garrison whose morale crumbled rapidly. And although the Germans acquitted themselves well in a bitter rearguard action right across the island, ending in a dramatic escape across the Straits of Messina, the fate of the island itself was really sealed from the moment the Allied landing forces got ashore. The plan had worked. After the war copies of all the letters were found in German Intelligence files. Although many of the *Abwehr* regarded them as suspicious, Hitler thought they were totally genuine, and as always it was his view which counted.

Paradoxically, all these deceptions made the greatest deception battle in the whole war—the preparation for the D-Day landings—progressively more difficult. Partly, it was a question of once bitten, twice shy. Every time the Germans were misled by a planted piece of information, the chances of their being hoodwinked by a similar plan again became more remote. Partly, it was a question of geography. As the war progressed, it became obvious that the Allies must make a landing in occupied Europe sooner or later. All the shoreline of France, Holland and Belgium was fortified and garrisoned with German troops. Wherever the Allied forces came ashore they would meet the Germans in strength, and good communications would make sure that reinforcements would reach the defenders more quickly than the attackers. Capturing ports, as the bloody and discouraging Dieppe raid had shown, was all but impossible against a strong and determined defence, and in fact the Germans had stepped up all their port defences as a result of that operation. If the Allies *did* succeed in taking a port, the Germans could easily make it unusable for months. Yet without a harbour, how could the Allies supply and reinforce their invasion troops quickly enough to stop the defenders pushing them back into the sea?

So began the biggest deception plan of all. On the face of it, northern France looked an impossibly tough proposition. The Germans had become prisoners of their own propaganda after the Dieppe raid. Although this had been intended as no more than a reconnaissance in force, a feasibility study of the problems of mounting an invasion of France, the Germans had trumpeted it to the world, and to themselves, as a full-scale landing which they had decisively defeated. The result was that they became over-confident of their ability to defeat the next attempt.

When forces earmarked for the invasion of North Africa sailed from Scotland, the troops had been issued with Arctic clothing, a fact picked up by German agents and radioed back to base, with the result that German forces in Norway went on to full alert. This was the beginning of a constant German preoccupation with the possibility of landings in Norway. Other rumours included the threat of landings in Spain or Portugal or both, or an invasion of the remoter parts of France itself. At the end of January, 1944, for example, the SS *Das Reich* Panzer Division, then resting and refitting in France after a severe battering on the Russian Front, was rushed to the Atlantic coast near Bordeaux to repel what was reported to be an invasion, but which turned out to be a group of Spanish sardine fishermen blown off course by bad weather.

Once the Allied invasion *had* been decided upon at the Teheran Conference of 1943, the old problem reappeared. Northern France was the shortest route to Germany, but also the most difficult. Landings in Norway might have been easier, since German troops there were much thinner on the ground, but communications were poor and landings there would have had to be followed by more landings in occupied Europe proper at a later date. All the theoretical advantages lay with a direct leap across the Channel from Dover to the Pas de Calais. This gave the shortest sea route, the closest air cover from airfields in Kent and Sussex, the shortest route on the other side to the heart of Germany, and had three major ports (Calais, Dunkirk and Boulogne). Lastly the whole area was the keystone of the coming V1 offensive against England.

Yet to be set against this was the one insuperable objection: the advantages were all too obvious. So obvious that the Germans filled the flat fields of Picardy and Artois and the adjacent areas of Belgium with more than 300,000 troops, ready to spring into action as soon as the first Allied soldier appeared on the beaches. Normandy, on the other hand, was a much poorer proposition from the Allied point of view, or so it appeared from the German side. There were no major ports between Cherbourg in the west and Le Havre in the east. Fearfully strong currents swept along the coast at every rise and fall of the tide; there were large areas of open beach which would have to be crossed, exposed to every change in the weather and every gun posted on the sand-dunes and clifftops. The sea crossing was three times longer, and once ashore there were hundreds of miles more of French territory to fight through before reaching the frontiers of Germany.

It was hardly surprising that orthodox German military opinion, led by Field-Marshal Gerd von Rundstedt, was firmly convinced that the Allies would come by the shortest and most obvious route, whatever the obstacles. It seemed as if the German commanders, having themselves profited so much by their clever use of the unexpected, were unwilling to allow their opponents similar originality. Only a few dissidents, led by Field-Marshal Rommel, were equally convinced the Allies would bank on the unexpected.

Ironically it was to be Rommel yet again who was the sharpest thorn in the side of the Allied plan. After the North African campaign, he had been placed in charge of the building and inspection of the anti-invasion defences of northern France, and he knew only too well how thin the shell of the much-vaunted West Wall defences really was. Knowing how much better the defences were in the Pas de Calais and the regions of the bigger ports than anywhere else along the coast, he assumed the Allies would realize this themselves, and would therefore attack the Normandy coast between Le Havre and Cherbourg, a region he knew was highly vulnerable. The German Seventh Army, which defended this coastline, was a nightmare hotchpotch of second-rate units and foreigners in German uniforms. Many units were made up of elderly soldiers or chronic invalids. There were twenty-three battalions of Russian prisoners along the Normandy coast. Their equipment was chaotic; between them they had ninety-two different types of guns, firing 252 different types and calibres of ammunition. Heavy artillery emplacements were few and far between; yet even these had to use twenty-eight different calibres of gun, many of them lacking even range-finders.

Rommel pestered the Army authorities for reinforcements for Normandy. He demanded the 21st Panzer Division. He demanded the Panzer Lehr Division, and he demanded the 12th Panzer Division, knowing that these were the only armoured formations strong enough to push the enemy back into the sea. But he also knew the Allies would use every scrap of their formidable air superiority to stop reinforcements getting through to aid a German counter-attack. For the Panzers to do their job quickly, he wanted them stationed as close as possible to the beaches – on the very clifftops if possible. There they would be almost immune from Allied fighter-bombers. He also demanded an entire corps of anti-aircraft guns from the *Luftwaffe*. Single-handed, he came close to stopping the Allied invasion plans without even knowing they had started.

Unfortunately for Rommel, his meteoric rise to fame and rank under Hitler's patronage had won him many opponents in the rigidly caste-conscious ranks of the General Staff. While Von Rundstedt and those who agreed with him had first call on the best units in France to reinforce the Fifteenth Army in the Pas de Calais, there were others who insisted on the few Panzer units allocated to the Seventh Army being kept well back from the coast, so as to be better able to deal with attacks over a much wider area. And as always, the man whose decisions counted most was Hitler, and *he* changed his mind from day to day. At one time he stood almost alone in agreeing with Rommel – and the tanks of the 21st Panzer Division rumbled through the streets of Caen, the first Panzers to be stationed so close to the Normandy coast since 1940. But then he changed his mind again, and while the Panzer Lehr

Division did move to Bayeux in the coastal zone, the 12th Panzer and the Flak Corps never arrived. By this time he was more worried about landings at Calais or in Spain, in Norway or even in Germany itself.

This was the situation which faced the deception unit formed in February, 1944, under the umbrella of the Supreme Headquarters, Allied Expeditionary Force (SHAEF). It was decided straightaway that the initial need was to build up a series of deceptions in tune with German prejudice and attitudes — since any deception works best when it reinforces what the enemy already believes. Since the Germans were so convinced that the Dieppe raid had been a full-scale attempt at invasion why not start by letting them think the Allies had lost heart about the whole idea?

One trump card in the Allied hand was the network of German agents which had been captured inside Britain. These were men whose fitness for their work contrasted oddly with German efficiency in other spheres: often they spoke indifferent English, had inadequate cover stories, were badly briefed on conditions inside Britain, especially over papers and other wartime restrictions. One by one, in quick succession, they were caught. Some were imprisoned, others put to death, but some who were both useful and co-operative were, in intelligence parlance, 'turned round'. In other words, they began to work for British intelligence, while sending messages back to their German paymasters; but the messages which were sent were exactly what the British wanted the Germans to know and to believe. They could hardly do otherwise, since they came from their own agents!

A young Spaniard, afterwards code-named 'Cato' by the Germans and 'Garbo' by the British, was so determined to work for the Allies that he volunteered while still living at home. Since the British showed no outward interest, he forced their hand by going to the Germans and getting himself recruited as a potential agent, and then going back to the British as a ready-made double agent. He was now a very valuable property indeed. A cover story was devised to transfer him to England, where with the aid of his intelligence bosses he was able to recruit a whole network of fictitious agents (known to the Germans as the Cato Orchestra), all of whom had convincing identities and cover stories which would allow them to feed titbits of information to the Germans when needed.

Garbo was the kingpin of all the double agents. In twenty months he sent the Germans 400 letters and 2,000 messages. During the months before the invasion, he was officially running a network of fourteen agents (many totally fictitious) and eleven contacts, including an agent in Canada and another watching Mountbatten at his headquarters in Ceylon. So highly did the Germans trust Garbo that they gave him a high-grade cipher for communication, with a regularly updated system of keys. This had the bonus of opening up many other German secret communications channels to Allied eavesdropping.

Oddly, the Germans trusted these agents implicitly, which seems especially strange in view of the German success in Operation Nordpol against the resistance in Holland. The capture of Allied agents dropped in Holland with radio transmitters had allowed the Germans counter-espionage organization to take over these transmitters and maintain direct contact with London. Before the fatal error was discovered, the Germans were able to send details for mass arms drops, and rendezvous instructions for more and more agents, all of whom were collected by the Germans as soon as they arrived. Yet this was essentially a defensive success, catastrophic as it was for those involved. Never did the Germans use their mastery of the Holland organization to change Allied policy or plans as part of an offensive strategy.

So the first of the deception plans began. Although the shadow of the Dieppe fiasco still cast a heavy pall over Allied hopes of a successful invasion before 1944, this was no reason to allow the Germans to rest content until that time came, and, during 1943, the first major cross-Channel deception, code-named Cockade, was mounted. Allied air activity, commando raids, broadcasts to the Resistance and other carefully dropped hints all hammered home the message of an invasion in the autumn of 1943. So worried would the Germans be by this threat that they would keep their troops in France without transferring units to deal with other Allied threats in Russia or in the Mediterranean area, or at least that was the intention. In fact, this particular programme failed completely. To begin with the Germans failed to react to some of its more subtle ploys, while the more obvious ones which they *did* notice were speedily dismissed for what they were. So confident were they that the cross-Channel invasion threat at that time was a deliberately inspired Allied myth that no less than five Panzer divisions and twenty-two infantry divisions were pulled out of France during 1943 to reinforce the fighting fronts elsewhere.

But Cockade, like Dieppe, had been a failure which paved the way for success by providing priceless experience. Yet, again like Dieppe, this experience had to be dearly bought: the promises of imminent invasion which had been intended for German ears had touched off Resistance attacks which brought savage, and effective, German reprisals. The result was that key networks collapsed, or were penetrated by *Abwehr* agents. Many agents were captured in the ensuing chaos, men who had been told details of the false invasion plans as if they had been the truth. After the war, there were bitter accusations of deliberate betrayal to further the deception plan – since the Germans were almost bound to believe detailed information extracted from trusted Allied agents under torture.

The most serious accusations centred on the so-called 'wireless game' manœuvres between the Allied and German secret intelligence services. If an agent was captured there were checks in the form of pre-arranged deliberate mistakes which had to be included in his signals back to London, so that he could prove he was not operating under German orders. If the checks were missing, London could either ignore the signals, or simply pretend not to notice, and use this channel to feed information to the Germans which would confuse or mislead them.

The problem was that this kind of communication was a two-edged weapon. If ever the Germans realized that London knew the agent was in German hands, then they might automatically treat any information passed through that channel as false, and by inverting everything they heard, thereby deduce the truth. So if this method was used, then every effort had to be made to convince the Germans that London still considered the agent to be genuine. In other words, if the Germans requested arms drops, then arms drops had to be made, on the basis that the weapons which were lost were a reasonable price to pay for the information thereby planted on the Germans.

Other, more disturbing, allegations were made after the war – that agents were sent in response to messages which were known to have been sent under German control and that these sacrificial agents were, unknown to themselves, given false information which they might well reveal under interrogation. These allegations were strongly denied, yet there were undeniably cases where agents risked torture and death to inform London that they had been caught and forced to operate under duress, and yet no action was apparently taken to stop the flow of arms and men from Britain. At the very least it smacked of incompetence; yet if it was part of a deliberate plan, it represented a brutally effective if cold-blooded way of deceiving the enemy on the largest possible scale.

Cockade's successor, Plan Jael, was much more carefully orchestrated and planned – but, most important of all, it was given more backing in terms of men and material than any of the earlier deceptions. From the beginning it was a world-wide scheme. Churchill, Roosevelt and Stalin had agreed during the 1943 Teheran Conference to produce a joint cover plan hinting at landings in Scandinavia and the eastern Mediterranean in an attempt to distract German eyes from the Channel.

Certainly, every hint of further operations in the south-eastern corner of Europe seemed to produce a violent German reaction. On the basis of information like that supplied by 'Cicero', valet to the British Ambassador in Ankara and later rumoured to be a British double-agent supplying the Germans with carefully chosen material, or reports like those saying that General Patton was about to lead the US 7th Army to attack Trieste, backed up by pictures of massed landing-craft concentrations (really being assembled for the Anzio landings), Hitler eventually reacted as the Allied commanders hoped he would. By the time the Anglo-American armies did land in France, he had sent a total of twenty-three infantry divisions and two Panzer divisions into the remotest reaches of Europe, more than a quarter of the total he had left to defend all of western Europe from invasion.

One of the agents reported back to Germany that labour troubles in the United States were holding up production of landing-craft. Another reported rumours that General Montgomery, notorious as a perfectionist in matters of preparation, was dissatisfied with the standard of training and had ordered the whole programme to be revised and started again. Yet another reported that the sites of embarkation camps in Kent had the hut foundations in place but that work on the huts themselves had stopped. Meanwhile Garbo himself sent his German control, General Kuhlenthal, details of an Allied plan for administering the liberated territories after the Germans had gone. This had been carefully written to give the impression that the Allied planners were hoping the Germans would pull out of their own accord, to concentrate on the growing Russian threat from the east. He was even able to send a copy of a leaflet to be distributed to the French people telling them what to do when the Germans left.

The idea of voluntary German withdrawal sounds like wishful thinking on a grand scale, yet British Intelligence already knew that Hitler had actually been toying with such an idea in August of 1943. If he withdrew from northern and southern Europe, he could massively reinforce his armies on the main eastern and western fronts, and he had actually ordered Armaments Minister Albert Speer to look into the economic effects of the plan. So the deception was right on target. At the same time, the London Controlling Section of the deception unit, under the command of Colonel Bevan, was ordered to concentrate on the apparent build-up of Allied strength in the Mediterranean to reinforce the no-invasion theory.

But the picture could only be maintained for a limited time. Before long the ban on wireless traffic, the level of which would have given the increased tempo of invasion preparations away to the German monitoring stations, would become a real obstacle to training. It would soon be necessary to assemble shipping and to carry out practice landings, all of which could not be hidden from German eyes. So the idea was changed to a progressive loss of Allied patience with waiting for the Germans to withdraw voluntarily. Now invasion was on again, although the Allied reluctance to attack northern France was still to be decisive. Now, said the next phase of the deception plan, the Allies were planning to land in Norway.

Hitler himself had been a firm believer in landings in Norway, as indeed he had been at one stage or another about almost every possible area. Now came new confirmation of his own intuition, against the insistence of his despised generals that the Allies would merely attempt the obvious. From Norway the Allies would cross to Denmark and be only a few hundred miles from the heart of the Reich.

For the Allies the benefits of the plan were twofold. If the Germans believed in a possible landing in Norway, it would tie up sixteen valuable divisions, including one Panzer division. At the same time any invasion preparations spotted in England would be less likely to lead to a build-up in Normandy.

This new deception went into action in March, 1944, but since each succeeding campaign was growing more and more complex, all the tiny details fed separately to the Germans had to add up to a coherent picture. The best way to do this was to plan the operation as if it really was going ahead; then all the details would be consistent. For example, when would the Norway landings ideally take place? Three weeks after the D-Day landings proper, since this would ensure that all units in Norway would still be there when the real invasion began. Where would the forces for Norway be assembled? In Scotland, which was the ideal jumping-off point for northern Norway, and in any case there were many units already there preparing for Normandy. Where would the landings take place? The detailed plan consisted of initial landings by two divisions in the north to capture Narvik, which would attract all the German forces to meet it. Then a second landing would capture Stavanger airfield. Backed up by commando units and by Russian air power through Sweden, the bridgehead would be expanded and two more divisions landed as reinforcements, with the objective of seizing Bergen and Trondheim and eventually Oslo. Once Norway was in Allied hands, it would provide a base for landings on the

9

Danish coast, the liberation of Denmark and the invasion of northern Germany.

This was exactly in tune with Hitler's own feelings; but drawing up plans was only half the battle. The Germans still had to be convinced that flesh and blood troops were being committed to carrying them out.

The plan demanded a total of eight divisions, based in Scotland or Northern Ireland and divided into three army corps, one for the northern landings and two for the follow-up landings in the south. And each of these units had to have an identity, a cover story, in many cases a real commanding officer and a convincing volume of radio traffic.

In some cases, real divisions could be used to back up the threat. There were two British infantry divisions in Scotland—the 3rd Infantry and the 52nd Lowland—making ready for Normandy, together with a Polish division and Dutch and Norwegian units. In Ulster there were the 2nd, 5th and 8th American Divisions and the British 55th Division. Added to these were fictitious units invented for the deception, many of which had been referred to in earlier messages so that the Germans accepted them as genuine. The 4th Army was the headquarters for the operation, backed up by the real 15th US Corps in Ulster and two fictitious British corps, the 2nd, which was given a 'headquarters' near Stirling and the 7th, which was located near Dundee.

Even when the 3rd Division had to be moved southwards as part of the real D-Day preparations, there was another division ready to replace it. This was the 58th Division, created as part of an earlier deception designed to reassure the Germans if reports leaked out about the number of experienced troops from North Africa now in London. The cover story was that small numbers had been sent home to help in the training of newly-raised divisions which would afterwards be ready for service in any theatre of war: the 58th was clearly one of these. It even had its own formation sign of a stag's antlers on a black square. The commander of 4th Army was General Sir Andrew Thorne, the GOC Scottish Command, a careful choice as his name was well known to the Germans from the years he spent as military attaché in Berlin before the war.

Now came the task of passing all this information to the Germans in small enough packages to appear convincing. In many ways the agents were the simplest, since they merely reported what they saw, or what they had been told to say. Yet double agents, fictitious agents or devoted volunteers, their messages had to be used with care—too much vital information from one source and the Germans might suspect a deliberate plant. So the cover stories were carefully drawn up. For example, one double agent code-named 'Brutus' was in fact a triple agent. Captain Roman Garby-Czerniawski, a Polish General Staff officer who was also a brilliant skier and a pilot, fled to Paris when the Nazis invaded his country. There he joined the Resistance and was eventually discovered and captured. The Germans wanted him to go to London as a spy on their behalf, in return for his own life and those of others in his network, but he refused. Finally, after the German invasion of Russia, he agreed. A contract was drawn up and Sergeant Hugo Bleicher, the ace *Abwehr* spy-catcher in France, was put in charge of arranging an escape convincing enough to fool the British. Bleicher and one of his men (whom Bleicher knew to be a double agent working for the Resistance) took Garby-Czerniawski for 'interrogation' to Paris. On the way a collision was staged with a *Wehrmacht* truck, which allowed the Pole to escape, at least convincingly enough to fool Bleicher's man that it was a genuine stroke of good luck.

Later Garby-Czerniawski rendezvous'd with Bleicher in secret, after which he was handed on to an Allied escape chain who spirited him over the Spanish border. When he reached London he reported to the British authorities, eager to work on their behalf at last. But they were reluctant to use him at first, not because they did not trust him but because the Germans might mistrust their own double agent, and might already be preparing to read his information in reverse.

He was ideally placed to visit Polish units which really were in Scotland, and as a genuine military man, his despatches could be expected to contain more detail than those of a civilian agent.

The campaign was begun, however, by one of the fictitious agents under the control of Cato. This agent was supposed to be an aide-de-camp to King Peter of Yugoslavia with pro-German sympathies. Because of his closeness to the monarch, the Germans saw nothing strange in his being able to visit command headquarters and the estates of the landed gentry where confidential war matters were apparently discussed quite freely. It was Garbo's fictitious ADC who set the scheme in motion by casually reporting a visit to the headquarters of the British 2nd Corps which had recently moved to Garter House, near Stirling, from Catterick in Yorkshire.

His second message revealed a lot more. Back in London briefly before visiting Northern Ireland, he said he had met an American officer at a party at the Cavendish Hotel, who was attached to the US 15th Corps in Northern Ireland, and who had invited the agent to call him up when he arrived there himself. In the course of their conversation, the American had let slip that the 2nd, 5th and 8th Divisions of his corps were all training there, alongside the 55th British Infantry Division. He even told the ADC proudly that the American troops had been complimented on their turnout by the 4th Army Commander on a recent visit. Coming from an ex-commander of the Brigade of Guards like General Thorne, that was really something!

All this information was signalled to the agent's controller in Lisbon for transmission to Germany. In the meantime, it was backed up by Brutus's visit to Polish commanders in Scotland, where he had discovered that the 4th Army HQ was at Edinburgh, and that it contained the British 2nd and 7th Corps under its command. At the same time, an imaginary Venezuelan from Garbo's organization had reported the presence of troops from the 52nd Lowland Division in Dundee along with others wearing a badge he did not recognize, that of the imaginary 58th Division in fact.

Brutus then reinforced this apparently independent information by ferreting out that the 4th Army contained an American Corps he had been unable to identify, and also that the 58th Division was based near Stirling, with a badge of a stag's head on a black square. He had located a Norwegian brigade at Callander (which was true); near Perth he had come across the Lowland Division, with its badge of an oblique white cross on a blue shield (also true). He had also heard of a British brigade in the Orkneys (the 113th Brigade was stationed there at the time), all of whom came under 4th Army. All these signs, in his opinion, pointed towards a possible invasion of Norway about the end of May; but he said this in a way which assumed the Germans already knew all about the operation, only the precise timing being in doubt.

The Venezuelan agent reported watching amphibious training operations where the troops had been burdened with Arctic clothing in spite of the warm weather. Wireless traffic, too, added to the picture. The supremely efficient German monitoring and direction-finding stations could usually pinpoint a transmitter to within five miles or so; therefore it was vital that this part of the operation appeared genuine too.

Overall commander of the 4th Army signals effort was Colonel Roderick MacLeod, with a team of some twenty officers helping him create the wireless traffic of a non-existent invasion army. Signals units representing individual formations were sent into remote parts of Scotland to act the part of units undergoing strenuous invasion training.

The wireless offensive opened on 24 March, 1944, just two days after Brutus's decisive messages. Many of the messages dealt with routine, boring matters of administration. Others simulated training exercises, artillery shoots, or waged their own brand of psychological warfare by reporting on tests of new weapons of unspecified type but terrifying performance. Others dropped carefully angled hints, sending detailed requisitions for Arctic clothing or enquiring what had happened to a consignment of skis. The rest was mainly gibberish, designed to give the impression of much important traffic in code, whose very volume would convince the Germans something big was happening in the area.

One genuine German spy was called in to play his part without realizing it. He had been recruited by the British to spy inside Germany, where the Germans had turned *him* round. Unfortunately, British Intelligence knew this, so that whenever they wanted to plant information on the Germans from a source they were certain to trust implicitly, they simply passed on the appropriate questions. In this case he was asked to send weekly reports on the density of traffic on the railway from Hamburg towards Denmark, to keep a careful watch on the Baltic ports and to monitor the frequency of German troop transports sailing for Norway.

Perhaps the most brilliant of all the individual facets of the deception plan was that described by Sefton Delmer in *The Counterfeit Spy*, his book on the Cato organization. An Austrian Jewish patents expert, code-named 'Hamlet' by the British, who had fled to Brussels after the Nazis had annexed his country and then been recruited by the *Abwehr*, had become another British volunteer agent. His British contacts had given him fictitious business representatives in London, whom he could then use to justify information which he could feed to his German controllers.

One of these London contacts was supposed to work for the insurance organization dealing with insured properties all over Europe. At the height of the Norwegian deception campaign, Hamlet reported to the *Abwehr* that his London agent had told him that the Ministry of Economic Warfare was searching his records for flour mills, bakeries, cold storage plants and printing workshops in Norway. Was there any operational significance in this? A few days later he passed on another letter (written by the deception team) saying that the Ministry had taken away, under the strictest secrecy, his files dealing with installations in Alesund, Bergen, Oslo, Narvik, Stavanger, Moss, Porsgrund, Kristiansand and Trondheim. This was a beautifully indirect way of telling the Germans the invasion would be split into two parts, one in the north and the larger one in the south.

Finally, Colonel Bevan of the London Controlling Section flew to Moscow by bomber to persuade the Russians to play their part of the deception, hinting at Russian supporting attacks through Sweden.

After much discussion, they agreed. They were also to try and persuade the Germans that the Red Army could mount no serious attacks on the Eastern Front before July (thereby lulling the Germans into a sense of false security before the next Russian onslaught then in preparation), and that instead Russian resources would be devoted to helping the Allied attack on Norway in the late spring of 1944.

By the beginning of May, the Russians were going into action. Russians ships and aircraft made obvious recces of the Norwegian coast, Russian wireless units simulated the build-up of an army in the Arctic, and a fleet was assembled in the Kola inlet, under the eyes of Germany's Finnish allies. Finally, Brutus reported the arrival of a Soviet military mission in Edinburgh, to liaise with General Thorne's headquarters for the forthcoming attack. The whole elaborate and ingenious trap was set at last. But would the Germans take the bait?

Swiss newspapers quoted German reports of rumours of an Allied invasion of Norway circulating in Sweden and Finland. *Luftwaffe* Intelligence reports spoke of Russian participation in the plans, and the consequent misgivings of many Norwegians in London. Ribbentrop's Foreign Ministry gave details passed on to them by the Swedes of increased Allied pressure for the right to use Swedish airfields. And German Intelligence passed on the complete details of the Anglo-American units and commands involved in the plan, together with the timings and probable objectives.

Yet in a sense the Norwegian deception was partly a failure. It was another scare for Hitler and his overworked staffs, so that it was another powerful blow to German morale. Yet it did not, as the Allies hoped, result in masses of men and tanks being sent from France to Norway. However good the deception, the sad truth from the German point of view was that there were simply no men who could now be spared. The commanders in Norway would have to meet the landings, when they came, with the troops they already had.

On the other hand, no troops moved in the opposite direction either. The invaluable Panzers, the coastal artillery and the thousands of infantrymen who could have played a vital part in the fighting in France were kept where they were until the real invasion was firmly ashore. And, paradoxically, the partial failure of the Norway deception paved the way for the total success of the most vital deception of all — convincing the Germans that the real invasion of France would come in the Pas de Calais rather than Normandy.

As the Germans grew more sceptical about the possibilities of an Anglo-American-Soviet joint operation in Norway, they began to see it as what it really was, a brilliant Allied deception. Yet having now assessed it at its true value, they went on to draw the wrong conclusions. They returned even more strongly to their own belief that the Allied landings would come through the Pas de Calais after all. For why, they argued to themselves, should the Allies bother to mount such a detailed and elaborate plan hinting at such a remote invasion target, unless it was to try to distract attention from a landing in the most obvious place of all?

This, though the Germans themselves never realized it until it was too late, was the concealed trap in the Allied plan—the trap of the double bluff. For the next stage was to take this German belief in the Pas de Calais and reinforce it by every means possible until it became a conviction so strong that they would go on believing implicitly in it in the face of all conflicting evidence. Later on, when the Allies were to plant the idea of landings in Normandy as a deliberate feint, the Germans snatched at the idea eagerly as confirmation of their beliefs. Faced with the daunting prospect of landing in an area with three times the density of defending forces of any other comparable stretches of coastline, what would be more natural than that the Allies should try any means in their power to persuade the Germans to disperse those forces to meet threats elsewhere? And what should the German reponse then be? To refuse to be deluded by the Allies' now notorious skill in deception, however convincing the evidence, and to keep their best troops and tanks where they were.

This was exactly the response the Allies wanted. Deception and reaction were to fit together perfectly. So in the late spring of 1944, with the invasion looming ever closer, the SHAEF team finally dropped the Norwegian hoax. They were now free for the biggest and most important challenge of the war. Of all the deception battles, this one had the highest stakes, and in a field where the real effects of success or failure are impossible to measure precisely, it produced the most amazing effects. At a time when every man counted, and when hours measured the difference between success or failure, their efforts were to play a crucial role in keeping an entire German army out of the fighting. While the fate of Europe swayed in the balance, eighteen crack divisions waited in the wrong place for six whole weeks, waited for an attack which existed only in the minds of the deceivers and of their victims in German Intelligence, until it was too late for them to take any decisive part in the fighting.

CHAPTER EIGHT

CALAIS OR NORMANDY—
THE *WEHRMACHT* DILEMMA

Now THAT everyone's attention, Allies and Germans alike, was focused on the northern coast of France, the situation became more difficult for the SHAEF deception team. Basically, the problem was this: of all the coastline from the northernmost tip of Holland to the western end of Brittany, there were just two possible sites for an invasion. There was the Pas de Calais, which possessed all the advantages, but which for this very reason was more heavily defended than anywhere else—and there was Normandy. With the Allies planning to land in Normandy, and the Germans confidently expecting them to come across the Straits of Dover, the task of the deception team seemed simple. But was it?

The main danger was trying too hard. If the Germans ever suspected that they were being given too much information about a Pas de Calais landing, their opinion could swing right round and opt for Normandy instead, with disastrous results for the invasion. If they erred on the side of caution instead, then the indications pointing towards Normandy might become more obvious anyway, so that the Germans could reach the same conclusion for different reasons. The balance was a fearfully delicate one, and it could still be upset by factors entirely outside Allied control, such as changes in the German hierarchy, or even switches in the direction of Hitler's strategic intuition

The major weight of opinion among the German generals, led by the Commander-in-Chief West, Field-Marshal Von Rundstedt, still tipped the scales heavily in the direction of the direct Pas de Calais assault. But the opposition party, led by Field-Marshal Rommel, favoured Normandy just as definitely. For the time being Von Rundstedt's faction held the most influence with Hitler.

The Norway deception had been code-named Fortitude North. The Pas de Calais deception was christened Fortitude South and was actually put into action before the Norway deception had been finally dropped. At the danger of confusing the Germans, the demise of Fortitude North gave an extra ring of truth to Fortitude South. The planners hoped the Germans would deduce that even the Allies would hardly mount two contradictory deceptions at the same time.

At the beginning of March, 1944, when most of the agents were sending back dramatic messages about the threat presented by Fortitude North and the attack on Norway to their German controllers, all that happened on the Fortitude South front was a visit by one of the Germans' own agents, a Polish officer with the code-name of 'Talleyrand' who had been recruited by the *Abwehr* while serving in Bucharest in 1940 and who had then been appointed assistant military attaché at the Polish Legation in London, to his controller in Lisbon. Talleyrand had been turned round by the British to such effect that he was now to become a vital link in the D-Day deception plan. And although his information was still concentrating on preparations for Fortitude North and the Norway invasion, one very significant snippet of information dealt with the identification of the British 34th Armoured Brigade which had now been tracked down to Kent — an ideal assembly area for a landing near Calais. Back went Talleyrand to England, with orders from his German masters to find out more about the Allied invasion plans, and soon.

When Fortitude South had first been planned, in Autumn, 1943, its objectives had been strictly limited. Once the landing actually took place in Normandy, then the Germans would know they had been duped and the Pas de Calais theory would be gone forever. Within hours of the landings the reserves would be on their way, and the deception plan would have done all it could to delay them. Officially, there were only six divisions left in England after the main force was ashore in France, and these would be nowhere near enough to maintain a real threat to the German defences elsewhere.

But this left out one vital factor — the paper divisions. As the Germans thrust deeper and deeper into Russia, Hitler had used his under-employed forces in the west as reinforcements. Whenever a new German attack was planned, or whenever a Russian offensive punched a hole in the German line, units were ordered from west to east at a moment's notice. Other units, badly battered in the Russian fighting, travelled west again to rest and refit, to replace lost equipment and train new recruits. Only the quietness of the French sector and the absence of an invasion threat made this possible, and only the threat presented by the paper divisions could change the situation.

Now a strange alliance was formed, between two parties on opposing sides, with no contact or communication, save a common objective for completely differing reasons. On the one side was Colonel Bevan and the London Controlling Section, and on the other side were the German Intelligence experts, anxious to highlight the importance of the work they were doing to a High Command which placed a very much lower value on Intelligence than on the Führer's intuition. Chief of them was Colonel the Baron Alexis von Roenne, the staff officer in charge of the Foreign Armies (West) Department of the German High Command's Intelligence organization. Von Roenne was a realist, and an expert – he was one of the first to see the Norway threat as a deception. He was appalled by the way in which the High Command was stripping the armies in France of their best units to meet the demands of the Russian front. So, from the best of motives, he played neatly into the hands of the Allied deception team. Every piece of information which came into his hands identifying new units in England was added to his appreciation of the Allied order of battle. And every new division added to the total strengthened his conviction of the folly of leaving France defenceless against the blow which he knew must fall soon. But his realism led him into the other trap – of believing that the real blow must fall in the Pas de Calais.

Roenne was a shrewd and able officer, and he had a remarkably accurate idea of true Allied capabilities at the time. Balkan rumours had no effect on his conviction that invasion would come across the Channel. One reason was that the highly efficient German *Y-Dienst* (radio direction-finding service) was constantly monitoring the level of radio signals traffic in different areas; even if the messages could not be read, the intensity of traffic provided valuable information in itself. And the *Y-Dienst* records showed that traffic in the Mediterranean was steadily decreasing, while signals traffic in Britain was growing.

There were other features noticed by the alert German experts. In particular, the radio nets known to belong to the crack US 82nd Airborne Division, a unit almost certain to be in the forefront of any invasion build-up, sounded false. These signals showed the unit to be still in Italy, yet the rhythms of the different operators and the patterns of transmission appeared to have altered. Could this be an Allied deception designed to hide the fact that the unit was already in England preparing for the invasion, while a few wireless transmitters had been left behind to conceal its departure?

Roenne ordered a special watch to be kept on all unexplained traffic picked up from sources in England. Sure enough there were new patterns detected in a code which could not be read, but which could easily fit the relocated division. Then at last came the piece of evidence for which they had been searching: a signal in plain language about a routine matter, a paternity case involving an Allied soldier — but it referred in passing to the 82nd Division command post at Banbury. From snippets of information like this, Roenne had already built up a remarkably accurate picture of the number of units in England.

Then events on the German side began to play directly into the hands of the LCS. One of Roenne's officers discovered that the *Sicherheitsdienst (SD)*, which ran a rival but far less efficient Intelligence service than the *Abwehr* which controlled Roenne's department, was now trying to belittle the *Abwehr's* contribution by constantly scaling down its estimates of Allied strength — partly because this was what Hitler liked to hear, and partly to cover up its own lack of information by branding the *Abwehr* evaluators as naïve and alarmist.

When Roenne realized that Hitler was using these false figures, reduced more often than not by as much as 50 per cent, as his justification for stripping the Channel defences to find reinforcements for the Russian front, he was horrified. How could the danger be averted? Hitler could never be made to believe that the *SD* was falsifying the figures — all that could be done, suggested one of Roenne's officers, was to feed him the true information indirectly. If the *SD* underestimated by half, then all Roenne's department had to do was double its estimates of Allied strength, and the figures then reaching Hitler would be something fairly close to the truth.

At first, the whole idea horrified Roenne. But, in fact, it was not even necessary to falsify the information. All Roenne's department had to do was count all the hints of new formations being fed to them by the Allied deception teams at face value, whether proven or otherwise. This would avoid the need to create deliberate fiction, and would also help maintain a picture which was always consistent.

This was a step in the right direction. The deception team's figures were now being accepted by the *Wehrmacht's* own Intelligence evaluators, even if they were being interfered with before they reached the High Command. Then came another stroke of luck. Hardly had Roenne's department begun issuing the higher figures than the *SD*, for reasons of its own, stopped scaling down the estimates! Suddenly, from May, 1944, onwards, the deception team's total of true and fictitious divisions, then standing at ninety plus, (of which only thirty-five really existed) became accepted as gospel.

Even when Roenne and *FHW* realized what had happened, it was not clear what could be done about it. To tell the truth was impossible. What had been done for the best of motives still smacked of inefficiency, or even downright treason. Reducing the total bit by bit was easier, but still fraught with difficulty, for would not the Allies be expected to try to hide units preparing for an invasion? And the psychological effect of any reduction in the apparent threat in the west would probably encourage Hitler to start sending more units to meet the mounting and all too real pressures in the east. So for all these reasons, *FHW* was stuck with its own inflated figures.

On the face of it, what von Roenne had been doing was good sense. As the Intelligence officer responsible for monitoring and assessing the Allied threat, he was erring on the side of caution. Identifying non-existent units was, after all, less deadly an error than failing to spot units which were actually ready and waiting to join the fight. And paper divisions could do no harm; only the units which actually existed could take part in the fighting.

It was an officer on Montgomery's staff who first spotted the limitless possibilities of the paper divisions, and how they could play as vital a part as any fighting unit. When General Eisenhower took over the Supreme Command of the Allied invasion forces, he put the planning of Fortitude South into the hands of Montgomery, Admiral Ramsay and ACM Leigh-Mallory; and when the staff of the joint commanders began to study the problem, it was Major-General Francis de Guingand, one-time chief of staff to Montgomery's Eighth Army, who proposed that the Normandy attack should not be revealed as the true invasion, even when the attack started. Instead, he suggested, why not build up the idea to the Germans of two separate army groups. When one actually turned up in Normandy, they would still be in doubt about the other. Was it intended as a reinforcement for the Normandy landings, or was it aimed at somewhere else altogether?

This was a master stroke. The continued non-appearance of the paper divisions in Normandy would strengthen German suspicions that another blow was due elsewhere. After all, what would be the point of the Allies holding back massive forces in England if Normandy was to be the only landing? And since the paper divisions amounted, together with other units which, though real, would not be committed to Normandy anyway, to a much larger force than that constituting the real invasion, then Normandy must be a feint, calculated to draw off the German reserves from other areas. What other areas? The bulk of the reserves were right there in the Pas de Calais, and all the German convictions that here was the only sensible place to invade were given new encouragement.

So planners of Fortitude South were given a new assignment. As the forces for the real invasion were built up in south-west England, they had to build up their paper forces in the south-east. This was the beginning of the myth of FUSAG, which stood for First US Army Group. Even before the invasion was launched, its role was vital. At any time, German reconnaissance or genuine agents might discover the build-up of forces in the south-west, which pointed inexorably towards landings in Normandy. Only if they also knew of much larger forces in south-eastern England could the Pas de Calais threat be seriously maintained.

This meant a greater effort to maintain realism than ever before, a gigantic theatrical performance entirely for the Germans' benefit, with no detail left uncovered which might crack the surface of the illusion. A huge signals effort went into action on 24 April. False unit transmitters were stationed in Kent, to be picked up and pin-pointed by the German direction-finder stations; and the genuine signals traffic from Montgomery's HQ in Hampshire was led by landline to Kent to be transmitted from there, thereby showing that the centre of the whole invasion effort was in the south-eastern tip of England. Air Force units in western and south-western England were also sending messages by landline to be transmitted elsewhere to make it appear they were based in southern and eastern England. At the same time aircraft broadcast messages to make it appear they were based in East Anglia.

Once again the information given by the radio intercepts had to tie in with the reports from the agents, and to make sure each scrap of information fitted into a coherent whole a complete timetable was drawn up for the dispatch of the fictitious units to their fictitious assembly areas in south-east England. Then the movements of the agents could be planned so that they were in the right place at the right time to fit the whereabouts of the units they were supposed to have seen. For example, the movement of the 11th Armoured Division from its base in Yorkshire to East Sussex was supposed to take a week, and the movements of agent Brutus were programmed so that he could be in Dorking to see it passing through, while another agent would spot it at its destination. Once the timetable showed that the unit had arrived, its transmitters would go on the air to add to the deception signals.

Other precautions had to be taken over real operations. All the preparations for landing in Normandy had to be repeated all along the French coast, lest the Germans measure activity sector by sector and draw the obvious conclusion. Teams of commandos in frogmen's suits were sent out by midget submarine to reconnoitre the invasion beaches —measure slopes, examine beach obstacles and take samples of the sand and clay. They were sent to cover all the other beaches along the French coast as well, in case of capture or any other leakage of information which might lay overmuch stress on Normandy. Later, when the pre-invasion air attacks began, badly needed aircraft had to be diverted to strafe the Pas de Calais area, where three sorties were made for every two on Normandy.

At the beginning of May Garbo passed on a message from one of his fictitious sub-agents that units of the American 6th Armoured Division had been seen in Ipswich. At this time, though the Allies had no way of knowing it, German aircraft had seen the build-up of troops in south-west England ready for the real invasion, and von Roenne's department was worrying about this apparent shift from the south-east. By a stroke of good fortune, Garbo's message, together with a series of follow-up reports, was to restore German faith in their original estimate. On 5 May von Roenne issued a situation report which showed two American divisions moving from the west to the south-east, although another agent's report had been misread to show reinforcement of the south-western areas, which could only mean Normandy.

More messages were sent as Fortitude South's timetable gained momentum. Another fictitious agent reported an American corps in the Folkestone area, together with more unidentified units. But on 12 May von Roenne reported that while the main forces were in south-west England, the growing number of divisions in the south-east might be used to mount feint attacks—exactly the opposite impression the Allies wanted to make.

A few days later he changed his mind again. Agent Brutus reported that the 4th US Armoured Division and the 20th US Corps HQ were located at Bury St Edmunds. Other messages showed the movement of the real 3rd British Infantry Division, previously used as part of Fortitude North, to the southern area. This, together with the move of 47 Brigade from York to Portsmouth, finally tipped the balance in the Germans' estimation in favour of the south-east as the main area of concentration. There was also much increased air activity by the RAF and the Americans off the south-east coast. In truth, this was to stop the Germans checking for themselves the presence of the fictitious units, but in German eyes it all seemed a convincing part of the great build-up.

Another German agent captured by the British—a Dane with the cover name of Hans Schmidt and the code-name 'Tate'—was given a different kind of cover to help fill in another gap in the picture. He had been parachuted into England at the end of August, 1940, and been caught almost immediately. By 1941 he was sending messages to his delighted German controllers, who had no suspicion that anything was wrong. He told them he had found a job as a farm worker in south-west England, which kept him free from conscription, and had even acquired a wife and baby son. The family part of the story was true enough, but in fact they lived in a London suburb. The cover story was chosen so that he could report that there were no signs of troop movements in his area.

Another British agent, a Frenchwoman of Russian extraction named Lily Serguiev, code-name 'Treasure', was based in Bristol. She was asked to pass on negative information to her German controller, Major Kliemann — no signs of large troop concentrations anywhere in the area. With the south-west so quiet, there could be nothing important planned for Normandy.

The time now came for Tate to play a more decisive role. The deception team invented for him a farmer friend at Wye in Kent, and the story ran that the kindly Tate would visit him from time to time to help with the farmwork at weekends. This gave him the pretext to see all sorts of useful sights on his journey and also to make friends with a fictitious railwayman at the local station who was only too happy to grumble about the upheaval caused by all the extra wartime traffic in the south-east. Soon Tate was able to pass on to his delighted German contacts the full timetable for the units of the First US Army Group, from their base to the ports from which they would sail for France.

Passing the Germans information through their own agents was fraught with risks, in case they should ever assume the agents had been turned and treat the information accordingly. Obviously, the ideal would be to convince one of their own senior officers, someone who could see for himself what was happening in England and then return to spread the word. This would corroborate the information which agents were sending in, and silence any doubts about their genuineness.

Throughout the war prisoners on both sides were sent home under the auspices of the Swiss Red Cross in cases of serious illness. In the late spring of 1944 General Hans Cremer, one-time commander of the Afrika Korps, became a candidate for repatriation. He was taken under the wing of the LCS, who arranged for him to be taken to London by road from his prison camp in South Wales. Officially it was let slip that he was being taken through southern and south-eastern England — but in reality his car followed a detour through south-western England, where Cremer could not help but see the huge build-up of all too real tanks, guns and men ready for the invasion. There were no signposts or place names to spoil the illusion. When he arrived in London he was

given dinner by General Patton, who was introduced as C-in-C of
FUSAG, and slips in the conversation dropped the name of Calais at
the right moment. By the time Cremer reached Germany on board a
neutral Swedish ship and reported to the General Staff, he was convinced that south-eastern England was swarming with men and weapons,
poised to land in the Pas de Calais.

By now the seeds had been firmly planted in the minds of German
military intelligence. In one way, it had been a close-run thing, with
the balance only being shifted in favour of the south-eastern concentration just three weeks before the invasion began. Yet in another way
it had left an awfully long gap during which the Germans might
discover the truth, so that all the careful work could have been wasted —
with awesome consequences on the success of the landings and the
course of the war. Planting the idea itself, as the planners now realized,
was rather less than half the battle. The major part now lay in preventing
the leakage of any information, however small, trivial or unforeseen,
which might discredit that picture in time for the Germans to take
remedial action.

So the period of the Great Scares began. First came the Affair of the
Chicago Parcel. In February, 1944, a badly-packed parcel had fallen
apart while being sorted in Chicago's Central Post Office. The astonished
post office workers had found it full of documents giving precise
details of the invasion, with places, units and timing. It was addressed
to a girl living in a part of the city with a large German population, and
sent to her by a brother working as a sergeant on Eisenhower's staff.
It turned out that he had addressed it to her by mistake, instead of
sending the documents to Transportation Division in Washington,
because at the time he was worried about her illness and too preoccupied
to notice what he was doing. Detailed security checks seemed to bear
out his story, unlikely as it seemed, and the planners could breathe
again. The next one came in the Channel on 26 April, when the
Americans ran a major exercise off the Devon coast to rehearse the
landing on Utah Beach. The sand-dunes behind Slapton Sands, a
long, wide beach between Torbay and Start Point, similar to the
relevant part of the Normandy coastline, had been studded with
replicas of the German fortifications, and the armada of landing ships
was manœuvring to attack it. It proved to be a fiasco, ample testimony
to how badly exercises like these were needed. It took too much time
to blow up obstacles, too much time to get tanks ashore, too much time
to clear a path for them to leave the open beach and strike inland.

But the worst was yet to come. After the exercise was over and its lessons were being digested by a gloomy staff, news came that some of the troops earmarked to take part in it had never arrived at all. A convoy of landing ships had lost contact with its escorting warships and been hit by a squadron of German E-boats, two of the landing ships having been sunk and a third badly damaged before the escort had returned and driven the Germans off. Ten of the officers on board the sunken craft were fully briefed on the invasion plan, while survivors said they had seen the E-boats cruising round among the men swimming in the water, possibly searching for officer prisoners. Desperately the staff ordered a thorough naval search of the whole of Lyme Bay, hundreds of square miles of open water, and one by one the bodies of the missing officers were found.

Still accidents went on happening. An officer left a briefcase full of invasion orders in a railway compartment, where fortunately it was found by a railwayman and kept under guard until security men could retrieve it. Papers fell out of an office window in Whitehall, and one of the copies could not be found. It was handed in two hours later by a civilian passer-by, having perhaps been seen by heaven knew how many eyes in the meantime.

Then there were the scares on the other side. Thanks to the network of French Resistance helpers throughout northern France, the Allies had up-to-date information on the movement of every German unit, a service which sometimes caused them more alarm than assurance. Starting in the first week of May, the ominous signs began to build up: the Panzer Lehr Division, one of Germany's strongest, best equipped and most experienced Panzer formations, was transferred from Hungary to join the army in France at Verdun. As the division was on its way, the orders were changed. Air reconnaissance sorties showed huge columns of tanks and vehicles on the move west of Paris.

One way of recording the precise movements and locations of every German division was through the French Resistance. Since it was the French who did the Germans' laundry, they were therefore given early warning of every move, and a new address for washing to be forwarded whenever the unit arrived at its new destination. The Resistance reported Panzer Lehr at Chartres and Le Mans only a hundred miles south-south-east of Caen and the British sector of the invasion beaches. At one stroke this more than doubled the armoured forces available to the Germans in Normandy, with the 12th SS Hitler Youth Panzer Division already grouped around Lisieux in eastern Normandy. Then the next blow fell: 21st Panzer Division, known to be stationed at Rennes in Brittany, itself only 150 miles from the beaches, was moving

10

forward too. Another of the toughest and most formidable tank units was approaching the beaches. Its headquarters was set up in Caen, less than ten miles from the coast.

Nor was this all. Three strong Panzer divisions within hours of the beaches was bad enough, where before there had been but one. Later that week the patient watchers of the Resistance reported that the 6th Parachute Regiment and the 91st Light Division were being switched to the Cherbourg area, at the western end of the invasion coast, and the next Panzer unit to arrive would be the 17th Panzer Division, ordered to move from its base at Poitiers 200 miles to the south. The Germans *must* know the Allied plans; yet could they be changed at this late stage, or were they committed to what must become a total failure?

Order, counter-order, disorder is an old and well-proven military maxim, and fortunately by this time it was too late to change the point of the invasion. All that the planners could do was wait and see, to discover how long this build-up would go on, and whether it really proved foreknowledge on the Germans' part, or whether they were simply random reinforcements made in ignorance of the Allied intentions. In the meantime the barrage of signals and messages of the Fortitude team poured across to the German Intelligence men, continually adding to the number of divisions now in Kent ready for the crossing to the Pas de Calais.

Though there was no way for the Allies to know, their fears were groundless. The endless differences of strategy which plagued Hitler's top field-marshals had produced a switch in policy. Von Rundstedt, the senior commander in the west, and a convinced Pas de Calais protagonist, wanted his reserve Panzer Divisions held well back from the front, so that it would be well placed to strike against any invasion attempt. Rommel knew better. He could see the logical reasons why the Allies should attack in the Pas de Calais, but, having been defeated by the unexpected at Alamein, he realized that the enemy were likely to rely on the value of surprise, just as they had in the desert. Now, as Inspector-General of the West Wall fortifications, he knew only too well how weakly Normandy was defended compared with the Calais area. It was only common sense to try and bring the one other probable invasion area up to something like the same readiness as the obvious landing site. But he differed from his colleagues on other points too. He knew the terrible effects of air attack on communications, which could prevent a strong central reserve more than a night's march from the beaches from ever doing its job. He wanted the Panzers overlooking the beaches, where they could drive the enemy infantry back into the sea before they could get their own tanks ashore.

The answer, as always, depended on Hitler, who alone could decide the siting of the Panzers. For a short time, during the first days of May, his intuition told him that Normandy — in particular the Cherbourg peninsula — would be the vital spot. For as long as it took Hitler to change his mind back to the Pas de Calais, Rommel had his way, and reinforcements began to pour in to plug the Normandy loophole. But he was soon disappointed, as Hitler swung back to Rundstedt's ideas again. Still Rommel resolved to win him over again, and resumed his lobbying with ever greater energy.

Fortunately for Fortitude South, German Intelligence had no such doubts. Every scrap of information, like the news that British agents had bought every copy of the Michelin map for the Calais area to be had in Geneva, or that Dover harbour was full of landing-craft whose range would only allow them to be used in the Calais-Boulogne area, was added to the picture now coming into ever sharper focus in Von Roenne's and Schellenberg's offices. Very few *Luftwaffe* reconnaissance planes dared cross the Channel in the teeth of Allied air superiority, but those which did were allowed to visit certain areas of Kent and Sussex, where the physical deception teams, under the direction of Colonel David Strangeways, had been busy since early April.

These used adaptations of the techniques which had been so useful in the desert, but on a much larger scale. Tanks, lorries, guns, armoured cars, half-tracks and tents were built out of wood and canvas or simply inflated rubber. More could be simulated by the use of semi-camouflage; a network of tracks leading to a large wood could hint at the presence of a whole regiment, yet all the markings outside what was really an empty copse had been done by a single lorry the night before. This was done sector by sector right round the south-eastern build-up areas, and judiciously shown off to the Germans — an enormous assignment, since the total of real units in the south-west demanded 163 new airfields, 125,000 hospital beds and 54,000 men, of whom no less than 4,500 were cooks, just to keep them supplied and fed.

So far the Germans thought they knew the place where the invasion would take place. What about the time? Once again they tried to read the Allied commanders' minds by proceeding along lines of the purest military logic. The attack, they thought, must come at dawn, and at high tide. Any other time must mean that the troops would have to cross hundreds of yards of open beach, a sitting target for the West Wall machine-gun emplacements. Add to this the need for settled weather and the heavy shipping movements in the Channel, plus the necessity for moonlit nights to coincide with the landing period, and the most likely date was around the middle of May. The *Kriegsmarine* plumped for 18 May and the Normandy coast. The Army knew better as far as the invasion site was concerned, but all units were alerted for that date; leave was cancelled and the German Armies in the west gritted their teeth and waited for the storm to break.

As 19 May dawned they were experiencing an understandable feeling of anti-climax. The Navy experts consulted their tide-tables and their weather statistics, and concluded that there was no more real danger of invasion before mid-August. In spite of the huge armies waiting across the water, the pressure was off for the time being, and reinforcements could proceed at a more leisurely pace. There was more time for leave, for training and for exercises like the war game to be held at Rennes at the beginning of June, when the commanders would fight a theoretical battle on maps based on the idea that the enemy would drop parachutists along the Normandy coast followed up by huge seaborne landings.

But by this time, the date was set. On 8 May, eleven days before the Germans expected the first troops to arrive, Eisenhower chose the invasion day. The Allied experts had studied the phases of the moon, the weather records and the tide tables; but they had, thanks to their frogmen, far closer knowledge of the German mines and beach obstacles than the defenders realized. To them the risks of crossing open beaches were infinitely preferable to losing ships and landing-craft, blown up and ripped open on underwater traps they could not even see. By reaching the beach at dawn with an exceptionally low tide, the troops would find Rommel's obstacles high and dry and comparatively harmless. With a rising moon just before the landing, the initial waves of parachute and glider troops could find their objectives, and, with the long days of midsummer, air and naval support could be used to the best effect. Only one obstacle remained: the early days of June were notoriously fickle, with unstable weather producing high winds and heavy seas, either of which could make the invasion impossible, or strand the first wave on the beaches without their support and supplies being able to follow them ashore.

But the date had to be settled, and far enough ahead to make it a calculation of probabilities rather than a weighing of real evidence. Eisenhower opted for 5 June with the 6th as a follow-up date and the 7th as a last resort before the timings all went wrong again. Until then, the huge mass of men involved in the landings would have to stay at their posts, all contact with the outside world cut off to prevent security leaks. And in the meantime the Germans had to be lulled into as persuasive a state of false security as possible. While information still had to be sent out of the Allied build-up, some means had to be found of showing that the landings could not possibly take place yet.

That means was found in the unlikely guise of an elderly lieutenant in the Army Pay Corps named Clifton James, who had appeared in a newspaper feature on account of his extraordinary likeness to General Montgomery. He was recruited by MI5 and trained to impersonate Montgomery to the life, although as an extra precaution he was told that he would be passing himself off as the General in England while the real Montgomery was in the Mediterranean. Only on the eve of his departure was he told that Montgomery would be fully involved in invasion preparations while Clifton James would be impersonating him on a carefully planned visit to Gibraltar, Algiers and a tour of North Africa, placed where he would certainly be observed by German agents — and the Germans knew well that if the invasion really was imminent, the one person certain to be on the spot and playing a major role would be General Montgomery.

At the same time, American generals were being shuffled around to suit the demands of the Fortitude deception. When General Omar Bradley arrived in England he was scheduled to take over command of the First United States Army Group, the real formation whose name had been borrowed for the fictitious army assembling in Kent. Lest leaks compromise the deception, Bradley's command was re-numbered the 12th US Army Group, and his senior and more notorious colleague, George S. Patton, named as the new commander of the 1st Army Group. This was sensible for three reasons: Patton was well known to the Germans for his dash and skill in mobile warfare, ideal for the thrust they believed would be developed from the Pas de Calais bridgehead. He was senior to Bradley, so this reflected the idea that the Pas de Calais army was the one chosen to deliver the decisive blow. And he was not scheduled to take over his real command in the Normandy bridgehead until a month after the invasion, so his presence there would not blow the cover story until then.

There was one snag: before all this had been decided, Garbo had made a reference to Bradley's appointment as provisional commander of the 1st US Army Group. Now a correction had to be issued. So another signal was sent explaining that Patton had now taken over and set up his headquarters at Wentworth. In the meantime, on the last day of May, Brutus sent a long message to Hamburg filling in the final gaps in German knowledge of Patton's new command.

So the hours ticked away. But as 4 June dawned, gales swept out of the west and the weather grew worse. Behind the Wall the Germans relaxed, knowing that invasion was impossible. Rommel left for Germany by car, to visit his wife on her birthday and also to see Hitler in another attempt to persuade him to send more Panzer formations to coastal positions. Preparations went ahead for the war-game at Rennes, and in the meantime a hundred miles to the northward, the long lines of ships for the D-Day Armada were already on their way, pounding into rising seas, rolling viciously in the troughs of the rollers as the wind and sea grew steadily worse. Soon the situation became all too obvious: to attempt the landing now would result in disaster. The ships had to be given the recall signal, whatever the danger of a breach of security, in the hope that the weather would change for the better by the next day. Gradually, in the midst of the storm, all the convoys turned back and retreated north. Had the Germans spotted anything? Miraculously, it seemed, no sign of the huge forces poised and waiting to strike had reached them.

But the truth was that the Germans knew full well what was happening—or they should have done. One of the loopholes in the Allied security net was its radio communications with the Resistance movements in Europe. Often these consisted of pre-arranged code messages to different groups to organize arms drops, pick up returning agents, or prepare the huge sabotage effort needed to support the invasion. The sentences were read over the BBC programmes after the news bulletins and gave no idea as to their real meaning. Some were snatches of poetry, others simple sentences like 'Jean did not shave this morning' or 'Aunt Marie sends love to her children'.

One of the last legacies which the *Abwehr* had been able to pass on to their colleagues in Army Intelligence had been a key to two of these messages, collected from one of their double-agents who had penetrated the Resistance networks. To give the French advance warning that landings were imminent, the BBC would transmit the first line of a poem by Paul Verlaine, '*Les sanglots longs des violins d'automne*'. This was to be a stand-by message, issued on the first day of the month in which the invasion would take place. The second line of the same poem, '*Blessent mon coeur d'une langueur monotone*', would follow as an alert, indicating that the invasion would begin forty-eight hours from the midnight following the time of transmission. Although the Germans could not be sure that this information was true, their listening stations monitored every one of the BBC transmissions, keeping an ear open for the fateful words.

Month after month had gone by, but the monitoring staff of Oberleutnant Hellmuth Meyer at Fifteenth Army headquarters at Tourcoing, near the Franco-Belgian border, had not heard the two lines of poetry. Not on 1 May, when the Germans had been convinced the invasion was almost upon them, nor on 17 May, when they believed it to be hours away, had either line been heard. Then, suddenly, after the nine o'clock news on the evening of the first day of June, there it was. '*Les sanglots longs des violins d'automne*', just as the *Abwehr* had said. Meyer's men redoubled their concentration, listening for the vital second line, and Fifteenth Army, guarding the Pas de Calais, was put on full alert.

Then just after darkness fell on the night of 3 June had come another message, a cable from the Associated Press offices in London: 'URGENT PRESS ASSOCIATED NYK FLASH EISENHOWER'S HQ ANNOUNCES ALLIED LANDINGS IN FRANCE'. It was incredible. There were no reports of any activity anywhere along the coast, and it was inconceivable that the Allies would announce an invasion over the radio before it had begun. Hour after hour the Germans waited anxiously, but by morning it was clear that it was just another false alarm. An urgent check on the Allied side revealed the reason: a teleprinter operator, bored and playing with the machine, had practised tapping out the most longed-for bulletin of all, without realizing the circuits were switched on and the message was going out over the air, to be picked up as far away as Berlin and even Moscow. A correction goes out soon afterwards, but the Fifteenth Army does not bother to react. They are still on full alert, as so often before during this invasion summer.

Finally, on the night of the fourth of June, at half past nine, Eisenhower's staff met at his headquarters at Southwick House near Portsmouth to listen to the latest weather reports from the meteorologists. Although their German opposite numbers predicted unbroken bad weather for several days, so definitely that many senior officers judged it safe to leave their headquarters for leave or meetings, the Allied weathermen had the benefit of more up-to-date information from out in the Atlantic. Now it seemed that a new weather front was approaching the Channel: the conditions would clear during the fifth and over the following night and the morning of the sixth the wind would drop to leave reasonably clear visibility and fair weather for just twenty-four hours, after which the weather would close in again. It might be just enough, but it was an agonizing decision to make. Within quarter of an hour Eisenhower had made it—the convoys could turn round again. The invasion of Europe had been postponed for just twenty-four hours. D-Day for the Normandy landings was to be the sixth of June.

Just over twenty-four hours later, at a quarter past ten on the evening of 5 June, two hundred miles to the eastward, Oberleutnant Meyer heard the words he had been waiting for, '*Blessent mon coeur d'une langueur monotone*'. Straight away the commander of the Fifteenth Army, General Hans von Salmuth, redoubled the alert and sent off priority signals to Von Rundstedt's headquarters at OB West, who alerted the OKW, the German High Command, who then alerted the rest of the armed forces in turn. Teleprinters chattered out the messages, as stage by stage the entire German military machine in western Europe woke up, and stood by on full alert—with one amazing exception. So firmly had Fortitude's deception buttressed the German commanders' own prejudices and preconceptions that everyone by now assumed the attack would begin in the Pas de Calais. No one at OB West, whose duty it was to pass on signals to all the units between Holland and south-western France, thought it worth adding Seventh Army to the list of units to be roused from sleep. And the slumbering Seventh Army was responsible for the stretch of the Normandy coast towards which, at that very moment, the greatest invasion fleet in recorded history was already moving.

CHAPTER NINE

THE GREATEST DECEPTION OF ALL

ONCE THE invasion was actually on its way the type of deception changed. For the time being, the long-term strategic aims of Fortitude South could take a back seat, although the programme, as we shall see, continued unabated through the most dangerous hours of the invasion, and resumed an importance which was to become even more vital afterwards. But as the fleet neared the coastline of northern France new deceptions were mounted with very different ends in view. Nothing could now be done to hide the fact that an invasion fleet was on its way. But before the Germans were able to see it with their own eyes, every possible trick would be used to disguise its approach, to exaggerate its size and mislead the enemy over its eventual destination. From the double-agent and the radio transmitter, the inflatable tanks and the aerial photographs, the emphasis now shifted to men with machine-guns and portable gramophones, to bombers loaded with strips of tinfoil and a squadron of naval patrol boats towing balloons on a wild-goose-chase up the Channel.

This enormous orchestration of deception began with the German radar system, which would give the first alarm the enemy could normally expect of the approach of the invasion fleet. Part of the pre-invasion air attacks had been directed against the radar sites. The huge Wassermann long-range early-warning radar array near Ostend had been crippled by rocket-firing Typhoons. Near misses had shaken the aerial-rotating mechanism out of true so that it was irretrievably jammed. The smaller but more numerous Mammut sites were even more vulnerable. Intelligence had discovered that each aerial carried a network of feeder cables at the back, and if these were cut by cannon and machine-gun fire from straffing fighter-bombers, then the whole system had to be completely re-calibrated in a long and complicated operation.

Some radar sites were bound to escape. So the invasion fleet itself was liberally provided with protection. More than 200 of the ships sailing towards the Normandy coast carried powerful jamming equipment, enough in concert to blot out the picture on every radar tube in the Bay of the Seine. But jamming these local radars in isolation was as good as telling the Germans exactly where and when to expect the invasion. Jamming every radar set in northern Europe would simply alert them everywhere; so the deception plan had to give them *some* information, which was wrong, rather than no information at all, which would merely trigger off a full-scale air and sea search.

A heavy force of bombers fitted with Mandrel jammers was sent up to blot out all the early-warning radars east of the Seine, right through the Pas de Calais area to the Belgian border and beyond. But two carefully planned gaps were left in the jamming coverage, intended to look as if caused by error or freak atmospheric conditions. Within these gaps the third and most important part of the radar deception was to take place.

The westernmost of these two gaps was the area of Cap d'Antifer, near Le Havre at the eastern end of the Normandy coast. Here there was a German Navy Seetakt long-range radar installation which could pick up shipping in the Channel. Also aloft on this night of invasion were the Lancasters of the RAF's élite precision-bombing squadrons, number 617 of dam-busting fame. But this time they were carrying nothing more lethal than Window, the strips of tinfoil which had been so effective in jamming the radar defences of the Reich itself. This time their task was not to attack, nor to jam, but to deceive, and this involved a standard of flying just as exact and even more vital than the night they blew holes in the Moehne and Eder dams.

The theory was simple enough: it was putting it into practice which created the problems. The Seetakt radar sent out a pulse 15 degrees wide, so that at ten miles range, it measured $2\frac{1}{2}$ miles across. At the same time, its perception of depth was limited to 250 yards, so that a convoy of ships sailing in tight formation would normally appear as a dense mass filling in the outline of the convoy. Building up this kind of picture with ships was out of the question, since hundreds would be needed, and none could be spared from the real invasion fleet. But aircraft could do the job, if they could drop Window in a dense enough and precise enough pattern to simulate the right kind of echo on the German screens.

The flying had to be done with absolute accuracy, and the pattern which the aircraft followed had to be carefully worked out to give the right results. It was decided to use eight aircraft in each wave, divided into two flights of four. Each flight would fly in line abreast on a course towards the coast, two miles apart from one another. After flying for eight miles, dropping Window steadily all the way, four aircraft would make a 90 degree turn to port, fly another two miles and then make another 90 degree turn to port, taking them on to a reciprocal course to their original one. They would then fly out from the French coast for seven miles, turn 90 degrees to port again, fly another two miles and then make a final 90 degree turn to port, which should bring them back over their original track and position them a mile nearer the French coast than their original starting point. At the same time the second flight would be following exactly the same pattern, but eight miles behind, so that there would always be one flight flying towards the French coast and another flying away.

This meant hairsbreadth flying of the most difficult kind. Any deviation from the pattern would distort the picture on the German screens and give the game away. But after two hours of this kind of flying, the performance of even the best crews could suffer. So both flights had to be relieved by another two flights coming out from England, who had to find their colleagues' positions and time their approach to take over the pattern at exactly the right spot without a break, and without any danger of collision in the pitch-black moonless sky over the Channel. And somehow they had to manage it not once, but three times, as the deception had to be maintained for eight hours in all.

Even the tinfoil strips created problems. Their length had to be related to the wavelengths of the German radar itself, and the long-range Seetakt had a much longer wavelength than the radars used against the night bomber offensive. So for this operation each strip had to be six feet long, impossible to handle in the cramped confines of a bomber fuselage. Instead, the strips had to be folded concertina-fashion, with weighted ends so that when they fell out into the slipstream they would stretch out to their full length.

In theory it would all fit perfectly. The Lancasters cruised at approximately three miles a minute, so that flying the set pattern and throwing out a bundle of Window every five seconds, would produce a huge echo measuring eight miles by eight and heading towards the French coast at a steady eight knots. The scheme was tried out against a British radar station of approximately the same type at Flamborough Head on the east coast, and it worked perfectly. So Operation Taxable, as it was called, was on. The 617 Squadron Lancasters were fitted with Gee, the navigational radio aid which had proved so helpful over Germany. Here it really had the best chance to do its job; close to its ground transmitting stations in England, it could give the crews of the bombers their position to within a few hundred yards throughout the long and difficult operation.

There was still one loophole. The Lancasters might fly the pattern perfectly, as indeed they did. The ominous pattern on the Seetakt radar might confuse and frighten the German radar controllers, as indeed it did. But the obvious next step was to send up German air-craft to investigate. Missing the ships, in darkness and bad weather, was all too easy on a visual search. But radar-equipped aircraft would be certain to pick up echoes from any fleet in the area, and their radars were shorter wavelength, shorter range but higher definition than the Seetakt sets, and they would miss the picture drawn so carefully by the orbiting Lancasters.

So the second phase of Operation Taxable swung into action. This involved a small fleet of naval motor launches, which had originally been shadowing the invasion fleet but, when it turned south for Normandy, had continued up-Channel, past Le Havre. Here six of the launches, accompanied by three mysterious converted air-sea rescue launches, headed for the French coast on the same course and at the same speed as the creeping radar picture sketched by the Lancasters above their heads. Each launch drew tethered behind it a sausage-shaped naval balloon, almost thirty feet long, called a Filbert. Each of these carried a nine-foot diameter radar reflector inside its envelope, equivalent on any radar screen to a ten-thousand-ton ship. The air-sea rescue launches carried an even more effective secret device called 'Moonshine', which picked up signals from the German radars in the area, amplified them and then sent them back as very large echoes indeed, providing ample and independent confirmation that there were indeed convoys of big ships off the French coast that night, heading very definitely *away* from Normandy.

Noises broadcast by the Taxable convoys included bosun's calls, anchor chains, bugles, loudhailer commands, squeaks of cranes and the noise of landing-craft engines. Warships, patrol aircraft and long-range guns were turned on the Taxable convoys, but not against the real Neptune fleet. One of the main reasons was that the Germans had carefully worked out a set of landing parameters which were quite wrong: they expected the Allies would only land at high tide, close to a major port in reasonably calm weather. But the landings came at half tide, they brought their ports with them and they took advantage of a lull in prolonged bad weather which the Germans, starved of weather information after the seizure and sinking of their weather trawlers, had no knowledge of.

But Operation Taxable was only half the convoy deception programme. Further east, in the heart of the Pas de Calais area, another mock convoy created by a Moonshine launch, eight patrol launches towing Filberts and the Stirlings of No 218 Squadron, flying a similar pattern to 617's Lancasters, made up Operation Glimmer. These aircraft used an even more sophisticated navigational aid, employing small radar transmitters and receivers on board the aircraft, in conjunction with a repeater station on the ground. This allowed several aircraft to use the system at the same time, which made life slightly easier.

While the bombers and patrol boats were courting the limelight to the eastward, deep in the radar fog off the Normandy coast the ships they were protecting were steering a very different course. Southwards they came, an awe-inspiring force of more than four thousand ships carrying the invasion army towards the beaches. Escorting them was a force of no less than 702 warships, from battleships to pound the long-range German gun batteries, to tiny minesweepers to clear a path for the fleet. There were even midget submarines, anchored off the French coast to guide the ships and landing-craft into the precise sectors of each beach assigned to them. And thanks to the cleverly-planned deception programmes, although many of the German forces were on special alert after the BBC code messages, not one member of the *Wehrmacht* yet knew anything of its presence.

It was vital to make the best use of this gift of surprise. Before the landings proper took place, parachute troops and glider-borne infantry would have to seize the causeways across land which had been flooded as part of Rommel's anti-invasion precautions. Other groups would have to hold a series of river and canal bridges at both ends of the beachheads, together with gun emplacements dominating the invasion beaches themselves.

The biggest problem was that all these objectives demanded quick action to begin with and heavy support to follow. Once the paratroops were established they had to clear landing-grounds for the gliders following them. Apart from more men, these could carry anti-tank weapons without which the parachutists would be powerless against the Panzers. Yet all their other objectives were exactly the places the Germans themselves would instantly reinforce in case of an attack. As soon as they realized what was happening over their heads, the *Wehrmacht* would rush men to every one of these objectives, and the parachutists would have to fight pitched battles against a wakeful and well-equipped enemy.

If, on the other hand, the Germans could be confused to the point that they failed to realize exactly which the danger points were, then their reactions might be slowed. Even the German Army didn't have enough men to guard every bridge and road-junction in the whole of Normandy — and this is where the next phase of the deception plan went into action. Since the presence of the paratroopers could not be hidden, the only way to keep the Germans guessing was to drench the whole of Normandy with paratroops, forcing the *Wehrmacht* to find out for itself which were real enough to matter.

So once again the dummy parachutists were recalled to the colours. Sticks of dummies were dropped from aircraft well away from the real landing areas, together with signal flares and strings of firecrackers to simulate the sound of a sudden sharp exchange between airborne forces and the defenders. Some were dropped to the south of the landing area proper, others on the coast, but most went down in the Fifteenth Army sector behind the Pas de Calais. Every extra drop was heard by someone, and contributed to the confusion now reigning at every German HQ in the area as rumour, counter-rumour, alarm and counter-alarm succeeded one another. At eleven minutes past two on D-Day morning, a telephone call came through for General Erich Marcks, commanding officer of 84th Corps, then preparing his maps for the fictitious parachute-drop-and-sea-borne-landing which would be the theme of the war-game he was due to attend at Rennes later that same day. It was from Major-General Richter, of the 716th Infantry Division stationed near Caen, reporting the British and Canadian paratroop landings in his area. Four minutes later calls went out to put the Seventh Army on full alert.

The confusion was total. Even when troops had been rushed to areas where dummy parachutists had landed, and found the evidence of trickery, a new element of doubt now began to creep into the reckoning. When new reports of parachutes and machine-gun fire came in from somewhere else, did this mean just another dummy drop, or did it warrant badly-needed men being sent miles to investigate?

There was worse to come. At other points far from the main landing areas, teams from the Special Air Service were being dropped to create more convincing illusions of battle. Equipped with fireworks and flares, they also carried portable record-players and loudspeakers. These were set up to give a complete sound picture of a battle, with shouts, screams, curses and explosions to go with the lights and flashes. SAS teams were dropped near Le Havre, between Lisieux and Evreux to the south-east of the invasion coast, and in a chain of drop-zones at the western end of the beachhead, from Lessay on the west coast of the Cotentin peninsula to Cerisy-la-Forêt to the east of St-Lô, and Villedieu-les-Poêles and Saint-Hilaire-de-Harcouët further to the south.

Distracting the Germans from the presence of the invasion fleet had been an enormous achievement. Keeping them guessing about the objectives of the parachutists had been another. But one of the biggest problems still remaining was protecting the huge and vulnerable air fleets taking the main forces of paratroops and gliders to secure the flanks of the beachheads. They were heavily escorted by radar-equipped Mosquito night-fighters, fast enough and deadly enough to make life difficult for any attacker. But this was very much a last line of defence. Even a single German night fighter finding its way into the stream of transport aircraft could create havoc and possibly wreck the whole operation. So a force of twenty-nine Stirlings and Halifaxes was sent to do another false attack. Over the open heathland east of Caen and around Cap d'Antifer they came, dropping Window by the ton to simulate the approach of another less well-guarded troop-carrying formation. As they reached the fake dropping-zones, they unloaded dummy parachutists, commando teams with flares, sound-effects equipment and firecrackers, to build up the picture of an even bigger battle.

Finally, to mop up the attentions of any other night-fighters left aloft that night, twenty-four Lancasters of 101 Squadron, backed up by five 214 Squadron Flying Fortresses, all from the 100 Group of electronic counter-measures specialists, feigned a massive bomber formation flying along the line of the Somme, dropping Window in the way used by normal RAF night attacks over Germany. The ground controllers would therefore have a vague idea of their whereabouts and could home night-fighters to attack them, but the precise location of each aircraft could then be blotted out by the Airborne Cigar jammer it carried, allowing the bomber's radio operator to find the German fighter-control frequency and then blot out communications altogether.

The result of this precisely-scored symphony of confusion was that the Germans took a disastrously long time to realize what was happening, and what it was they were dealing with. The Taxable fake convoy attracted reconnaissance aircraft sure enough; time and again the Moonshine equipment had picked up radar pulses and dutifully fed them back into the dark sky overhead. The Glimmer convoy attracted even more attention; once the reconnaissance aircraft and the coastal radars had agreed on its position, the huge radar-controlled coastal batteries at Cap Gris-Nez opened up, pouring tons of high explosive into the almost empty Channel. A naval alert then went out, and a squadron of fast patrol boats was sent to the area, but by then the fake convoys had reached their target areas ten miles off the French coast. The naval patrol launches played recordings of rattles and clanks and splashes characteristic of large ships anchoring offshore; then they departed as silently as they could under the cover of a smokescreen.

Reports were coming in from all over Normandy of parachute landings and aircraft sightings, as they were in the Fifteenth Army area too. But the night fighters were missing. All those who had taken off were homed in on the bomber formation put up by 100 Group; but once they entered the area covered by the bombers' ABC jammers, their controllers heard not another word from them until the night was over. Likewise, the fighters themselves were virtually deaf and blind. Only one of the Lancasters was lost, and even then its crew was saved. So what could easily have been aerial carnage was an entirely bloodless victory for the Allies.

At three in the morning, when the parachute troops were well on the way to capturing most of their objectives, Major-General Max Pemsel, chief of staff of the Seventh Army, called his opposite number at Rommel's headquarters, Major-General Hans Speidel, to tell him that it must be the invasion. He also passed on a report from the German naval units at Cherbourg that the noises of ships' propellors had been picked up in the bay of the Seine off the Normandy coast. Speidel for his part believed the attacks were still a local operation, designed as a diversion for the invasion proper. And when von Rundstedt's headquarters rang up for a report, Speidel thought it possible that the reports of parachutists were nothing more than baled-out bomber crews shot down during the air attacks over northern France. Thus reassured, Rundstedt's men saw nothing to disturb their view that the main blow was still to fall in their own Fifteenth Army area. All they, and the rest of the German Army, could do was wait until the situation became clearer.

By five that morning the situation *was* clearer. Major-General Pemsel was still continuing his one-man crusade to force the army to admit that an invasion was on its doorstep. More and more reports of ships' engine noises were coming in all along the radarless, fogbound coastline, and even Von Rundstedt decided it was time to act. Officially only Hitler could release the Panzer Divisions which would be vital in repulsing any landing; but the wily general ordered the 12th SS Panzer Division and the Panzer Lehr Division to start moving towards the coast before he sent a message to Hitler's headquarters asking for them

to be put under his command. Which was just as well, for when the
message reached the High Command, Hitler had retired to bed an hour
before with a heavy sleeping-draught and was not to wake for another
five hours. But Von Rundstedt was only being careful. Moving these
divisions forward did not mean he accepted this was the true invasion.
It was simply the kind of precaution which a professional and prudent
officer would take. The far more powerful and more numerous units of
the Fifteenth Army stayed alert, but they stayed where they were.

In some ways the deception had already succeeded better than many
of its most enthusiastic backers had dared hope. In the vital programme of beach defences most of the effort had gone into the coast
opposite the Straits of Dover, and the Normandy beaches were still a
long way short of what Rommel had regarded as the absolute minimum
in tank-traps, minefields and landing-craft obstacles. The German
Navy had developed the deadly pressure mine which was almost
impossible to sweep, and which could have caused catastrophic losses
among the invasion fleet, but the first ones available had been laid
between the Seine and the Belgian harbours, leaving the Normandy
coast unguarded. Fuel supplies had been concentrated where most of
the army was stationed—on the wrong side of the Seine, where it
stayed when the blowing of the bridges and the increase in Allied air
activity cut the road and rail links into Normandy. The day before the
invasion *Jagdgeschwader 26*, the *Luftwaffe*'s last fighter wing in the
area, had been moved to eastern France. Only two FW190 fighters
were left in the invasion area on D-Day, and those German bombers
which did attack the beaches had to run the gauntlet of Allied fighters
with escort.

The deception team were loath to abandon their cover story now,
when the Germans still seemed as convinced as ever of its truth. A
month before, Carlos Reid of MI5, Garbo's case officer, had realized
that the Germans would be in the most vulnerable situation after the
landings had taken place, not knowing how much of their forces to risk
on repelling the attack and how much would be needed to counter other
blows. What chance was there of producing some really dramatic
gesture by their agents, which could convince the Germans more than
ever that these were men to be listened to? Men, moreover, whose view
of the situation would show them how best to cope with the situation
arising out of the Allied landings?

Reid's answer was to have Garbo send the Germans the details they
had wanted for so long—the time and place of the Normandy landings—
before the landings actually took place, but too late for them to take

any active precautions. With the paratroop landings in full swing and the first seaborne troops due ashore at half past six in the morning, Garbo warned his German controller to stand by for a message in answer to an earlier question about the whereabouts of a particular unit—the reason being that he was awaiting a report from one of his sub-agents—and that the message would be sent in code at 2.30 on the morning of 6 June.

The cover story was that Garbo would then call his German controller with full details of the invasion forces—where they were heading and which units were taking part—information which had come into his hands through his network of agents. But the Germans stayed off the air for six hours. By the time the message did get through the troops were ashore, but the deception team was able to make up for some of the delay by making the message even more detailed. Later that day Garbo was able to send full details, by virtue of his cover job at the Ministry of Information, of a directive ordering all speculation about or mentions of 'diversions' or 'alternative assault areas' to be avoided. He also quoted draft speeches being prepared for Churchill and Eisenhower to give to the French people where several mentions were made of the need to avoid acting prematurely, and references were made to the 'opening phase' or the 'first assault' of the campaign and the great battles which 'lie ahead'. The following day he reported a furore over Churchill having used the phrase 'the first of a series of landings in force upon the European continent' in the House of Commons, which it was feared would alert the Germans to the main blow of the invasion.

All this information had the ring of truth to the bemused Germans. Backed up by Garbo's reports from his sub-agents in Scotland and the south-east that the major group of divisions were still firmly in position, they strengthened the hand of the Pas de Calais wait-and-see school against the increasingly vocal Normandy here-and-now protagonists. Later on that first day of the invasion Colonel Von Roenne summed up all their convictions in a detailed situation report on the landings in which he re-emphasized that the Normandy operation, although large in scale, still only comprised a relatively small part of the troops available. He referred to the 'sixty large formations held in Southern Britain' of which 'only ten or twelve divisions . . . appear to be participating so far'.

Roenne went on to stress that 'further operations are planned' and referred to 'statements to that effect by Churchill and Eisenhower'. He pointed out that the forces in southern England were held in two

army groups, the 21st (British) Army Group in the south-west, commanded by Montgomery, and the 1st (US) Army Group in the south-
east, commanded by Patton or Bradley. Roenne explained that all the
forces so far identified in the bridgehead had come from areas west of a
line between Brighton and Oxford, and cited official announcements
that Montgomery was in command of the actual landing forces. But the
crux of his report, the key to the thinking of most of the German
commanders in northern France, lay in two especially significant
paragraphs:

'Not a single unit of the 1st United States Army Group, which
comprises around twenty-five large formations north and south of the
Thames, had so far been committed. The same is true of the ten to
twelve combat formations stationed in central England and Scotland.

'This suggests that the enemy is planning a further large-scale
operation in the Channel area, which one would expect to be aimed
at a coastal sector in the central Channel region (i.e. in the Pas de
Calais).'

So far, then, the Germans had stayed firmly on the hook baited so
carefully by the Fortitude South planners. But, whether or not they
stayed entrapped, the agents could still be used to advantage in the
tactical as well as the strategic field. Already the development of the
beachhead was emerging. While the British and Canadian forces were
defending themselves against the increasing pressure of the local
Panzer forces in the open tank country around Caen, the Americans in
the cramped bocage to the west were preparing for the great breakout
which was to seal the invasion's success two months later.

Another deception involved the scheme for supplying the armies with
petrol once they were ashore in France. The ingenious PLUTO
(Pipeline Under The Ocean) was the key, but if the Germans had any
knowledge of the plan, then all they would have to do was look for the
shore installations in England to know exactly where the main thrust of
the invasion would be aimed. So more deceptions were arranged: the
real pumping stations were disguised as dummy seaside bungalows
amid rows of real ones, while at Dover—ideal site for a non-existent
pipeline aimed at the Pas de Calais—an elaborate fake oil terminal was
built from wood, canvas and old sewer pipes. Clouds of dust were
raised to show work going on at a furious pace: Montgomery, Eisenhower and King George VI visited the site and spoke to the workers,
while a celebration dinner was held to mark the site's completion, all
these events being faithfully reported in the newsreels. The whole
complex was heavily protected by fighter patrols, and so tempting did it

appear to the Germans that it was shelled several times by their long-range guns at Cap Gris Nez, after each of which fake fires were lit to convince them of their success.

On the first day of the invasion it was obvious that the Americans at Omaha Beach had encountered the strongest part of the defence system. They had had heavy losses on the beach and had to surmount grave difficulties in breaking out inland. Montgomery wanted some titbit of information fed to the Germans which would draw more of these defending forces against the British and Canadian sector, giving the American Fifth Corps a chance to extend and consolidate its beach-head.

So on 8 June Garbo sent another message, reporting that one of his agents had discovered that the 3rd British Infantry Division was on its way to the invasion, and was being followed by the Guards Armoured Division. The Germans had already identified the 3rd Division in the battle, so they had every reason to expect the Guards Armoured at any moment; and that same day Roenne's unit warned the German forces of its imminent arrival.

But as another two days passed it seemed that at long last even the most dyed-in-the-wool opponents of the Normandy landings as the most serious threat were beginning to change their minds. One of the chief arguments — apart from information and messages from England — in favour of the idea of Normandy as a diversion had been the absence of a harbour within easy reach of the beachhead. But even before the landings had taken place, German aerial reconnaissance had spotted the huge concrete caissons intended for the Mulberry artificial harbour to be built at Arromanches. These were moored off Selsey Bill, and it was a fairly obvious deduction that they would be used to construct some kind of harbour or dry-dock. And the squadrons of old battleships and merchantmen, sixty in all, to be used as blockships for the outer breakwaters, had sailed in the small hours of the invasion morning. On the following day, they were being manœuvred into position off the beaches, and it was fairly clear what they were being used for, posing the question — would the Allies waste all this material and ingenuity on a mere deception?

So after two days, with the Allies still firmly ashore in Normandy, Rommel finally succeeded in convincing von Rundstedt of the serious-ness of the invasion, and von Rundstedt and Rommel between them set to work on Hitler. Finally, on 8 June, after seeing his orders for the speedy elimination of the bridgehead disobeyed because the forces on the spot were not strong enough to do it, Hitler changed his mind.

Orders went out to the Fifteenth Army to detach troops as reinforcements for the units hard-pressed by the increasing Allied strength in
Normandy. This time no half-hearted measures would do, and the
action which the Allies had been dreading ever since the invasion plan
was settled began. No less than five infantry divisions were put under
orders to move to the Seine bridges, together with two Panzer
divisions, the crack 2nd Panzer and 1st SS Panzer Division, equipped
with Panther and Tiger tanks which could easily out-gun the Allied
Shermans and Cromwells.

All the same, the Germans still needed all the information they could
get, and Colonel Meyer-Detring, von Rundstedt's Intelligence chief,
sent Garbo's controller, Kuhlenthal, a request for any more messages
like the Guards Armoured Division one of 8 June. This was the
deception team's last chance to keep at least some doubts in the minds
of the Germans. They replied, through Garbo, with a marathon message
—it took more than two hours to transmit in code. It was no less than a
complete rundown of the whole Fortitude deception plan, presented as
Garbo's own summary of what the Allies were up to.

For the Germans, it made electrifying listening, ample confirmation
of their own worst fears, now borne out in detail by a totally reliable
agent with a recent scoop on the landings in Normandy already to his
credit, and a network of agents working under his orders in the heart of
the enemy base areas. He told them there was no doubt that the
present assault was intended as a diversion, and that its very strength
and ingenuity were part of the trap. Only if the vital German reserves
could be enticed into Normandy could the main Allied blow be directed
elsewhere with some chance of success. Garbo emphasized that a
'massive concentration' of forces in south-eastern England was still
remaining in waiting. As in Roenne's summing up of two days before,
two paragraphs give the key to his message:

'The constant aerial bombardment which the sector of the Pas de
Calais has been undergoing and the disposition of the enemy forces
would indicate the imminence of an assault in this region which offers
the shortest route to the final objective of the Anglo-American illusions:
Berlin. An assault here could be supported by constant bombardment
from the air. The bases of the enemy air fleets would be conveniently
close to the battle area. They would then fly their attacks in the rear of
our forces facing the enemy landing in the West of France. I learned
yesterday at the Ministry of Information that there were seventy-five
divisions in this country before the present assault began. Supposing
they were to use twenty to twenty-five divisions on the present assault,

they would still have fifty divisions available for the second strike.

'I trust you will submit my reports for urgent consideration by our High Command. Moments may be decisive at the present time. Before they take a false step through lack of full knowledge of the facts, they ought to have at their disposal all the present information. I transmit this report with the conviction that the present assault is a trap set with the purpose of making us move all our reserves in a rushed strategic re-disposition which we would later regret.'

It was a masterpiece: it contained something for everybody. It pre-empted the arguments of Rommel, since the stronger the Allied build-up in Normandy seemed, the more Garbo's words were borne out. It flattered the Generals, since it spoke of the danger of their making a wrong move through lack of information, rather than lack of judgment. Now they had definite information that the main blow would fall in the Pas de Calais, so there was no excuse for them making any more mistakes. And it had something for the man whose orders counted more than anyone: Adolf Hitler. Threats of Allied strength, only waiting for the very action which they were now about to take to be unleashed in their rear, would work with the Generals, but not with the Führer. He would not take the possibility of an Anglo-American advance on Berlin seriously, even at this stage of the war. But the idea of doing just what the Allied planners thought he would do must have been intolerable to the supreme warlord. Now he had the chance to outwit them, to prove once again that he was a match for any enemy generals and planners. It was irresistible.

Garbo's message was completed in the early hours of 9 June. By dawn the following day orders were on their way from the Führer's Headquarters to von Rundstedt — and at half past seven in the morning the commanding officer of the 1st SS Panzer Division received a message instructing him to stop preparations for moving to the aid of the Seventh Army and to stay where he was until further notice. Similar orders went to five of the other six divisions under notice to move.

The deception had worked again. Amazingly, as the Normandy bridgehead continued to expand, the myth of the Pas de Calais continued to exert its grip on the German commanders. On 12 June, six days after D-Day, the V1 offensive began, but even the flying bombs falling on London failed to produce the main attack or any doubts in the minds of the German generals. Although the 2nd Panzer Division was sent from Fifteenth Army to Normandy on 10 June, it was the only unit to move, out of the original seven, and by then one division, however powerful, could make little difference. Allied advantages in men and material, above all the effects of Allied airpower, were beginning to make themselves felt.

Eight days later there were twenty Allied divisions in the bridgehead, as forecast in Garbo's message to the Germans of ten days before – but they were opposed by the equivalent of only a dozen German divisions, another six having already been wiped out in bitter fighting. Reinforcements were being dredged up from any and every source to plug the gaps which the Allies were tearing in the German line – any source that is except for the strong and well-equipped units in Fifteenth Army. The 2nd SS Panzer Division was called up all the way from Toulouse. It had to fight for every mile against the Resistance. The 17th SS Panzer Division, resting after return from Russia, was brought up from south-western France, even though it had no tanks left at all. Some regiments travelled in requisitioned French buses, while others rode bicycles or marched on foot 200 miles to the battle area.

On 26 June the Americans captured Cherbourg and began to push southwards towards the base of the thinly-held Cotentin peninsula – but still the Fifteenth Army waited for the 'main assault'. All Hitler would agree to was the despatch of the 1st SS Panzer Division on its long-delayed journey to Normandy, in company with two other Panzer units, the 9th and 10th SS Panzer Divisions, which had been withdrawn at great risk from the eastern front. They were ordered to hammer through the British positions and drive to the sea. But to reach the attack area 1st SS Panzers had to detour all the way south to Paris and then back to Normandy to find an intact bridge across the Seine; *en route* it was cut to pieces by Typhoon fighter-bombers with tank-piercing rockets. The attack, when it was finally launched at the end of June, failed. Not least because another Panzer division which could have joined in had to be sent instead to fill the gap left by the 1st SS Panzers in Fifteenth Army, so that that still inactive formation should not be prevented from its deterrent role.

After this the Fortitude team were able to write their own script. As the few real units used as the basis for the First US Army Group were sent to Normandy, they were replaced by newly-invented formations. But the biggest problem was the man at the top – Patton himself. He was becoming increasingly restive at his non-operational command and demanding a place in the real invasion army. Yet he had been such a splendid choice as the figurehead of the deception that it was now an almost insuperable problem to find someone big enough to succeed him without the Germans concluding that the south-east Force had declined in importance.

In the end Eisenhower was prevailed upon to contact General Marshall in America to find a suitably renowned successor to take Patton's place. They sent Lieutenant-General McNair, Commander-in-Chief of the ground forces in the United States, and he arrived not a moment too soon, since Patton himself had actually been in Normandy since 9 July. The Germans themselves got wind of the news a few days later, and though Garbo sent cover stories through his sub-agents explaining the switch, this involved moving several divisions over with him, thereby weakening the forces supposedly still waiting to deliver the main attack.

With every day that passed, Fortitude became harder to sustain. As the Normandy beachhead continued to expand, as more and more first-line Allied fighting units were identified, and as day after day passed without the landing barges appearing off Calais, it became more and more obvious that the invasion which the Germans were already fighting was indeed the decisive one. On the last day of June, Hitler began planning for a massive Panzer strike against the beach-head. He too was beginning to conclude that the Pas de Calais landing could now only be a secondary operation, if it ever took place, and the best way to smash the Allies in Normandy was, as always, to use the waiting divisions of the Fifteenth Army.

Somehow the Allies had to keep the Fifteenth Army where it was. The Germans must have noticed that, if landings in the Pas de Calais *were* coming, 6, 7 and 8 July came closest to repeating the conditions which prevailed on the original D-Day. So, late in June, naval forces began shooting up the 'invasion' coast. Commando raids were stepped up, and a force of 2,000 bombers with fighter escort began pounding the whole Fifteenth Army area as a softening-up operation.

Still the German preparations went on. Rommel, as commander of the counter-offensive, was to use nine whole Panzer divisions and one infantry division, from a variety of sources. Now Hitler gave him the 116th Panzer Division and six more infantry divisions from the Fifteenth Army.

Only on 21 July did von Roenne admit that the diminishing of the forces known to be left in England made a separate landing less likely. Six days later he admitted it no longer had 'the capacity to carry out large-scale operations' and that it 'no longer constituted an active danger'. Yet this was no case of the awful truth suddenly dawning on the deluded victims. Rather they tended to believe that the un-expected success of the Normandy landings had resulted in the Allies changing their plans. Instead of running the risk of another landing against heavy defences, they had merely backed the favourite and poured all their efforts into what had been intended as a diversion, but which was going unexpectedly well.

Even then there were no violent changes of plan on the German side. Furious counter-attacks against the British positions above Caen by all but two of the nine Panzer divisions and all four of the heavy tank battalions in Normandy failed to push the Allies back, but British and Canadian attempts in the other direction were just as unsuccessful. Yet all this time the Americans in the west were building up their strength for the breakout. On 26 July they began to move, with an aerial bombardment which totally destroyed the Panzer Lehr Division lying in their path. A mighty flood of troops, tanks and guns from Patton's 3rd Army began the dash through the narrow gap near Avranches, fanning out to begin the race which would end in the liberation of most of France west of the Seine.

The American breakthrough, in the eyes of the Führer, made Operation Lüttich. the German counter-attack, even more necessary. Kluge* was advised to attack westward to cut off the advancing American spearheads. Despite his very real doubts, he complied, his troops bursting in on the surprised Americans without a single round of artillery fire to give warning of their approach. Although Ultra had once again told the Allied commanders what to expect, the warning had not filtered down to the units directly involved.

Yet the Americans fought back with unquenchable spirit, and the effect was the slowing, blunting and eventual stopping of the German thrust in its tracks. For Kluge, this was exactly what he had expected; the last straw was a furious Canadian armoured attack on his northern flank. Kluge called off the attack in the west, and was sacked by Hitler, who took over on his own account, determined to force Lüttich through to final victory.

This gave the Allies a precious new opportunity. As long as Hitler kept the German forces attacking down their long and vulnerable salient, there was a chance that the American forces which had already broken out, could swing round to attack the Germans' southern flank and catch the *Wehrmacht*'s irreplaceable attack force in an enormous trap.

So deception was employed once again. Agents like Garbo and Brutus reported that the Americans were running into trouble in Brittany, and were having to withdraw troops from Normandy to

---

* Field-Marshal Kluge had replaced Rommel after the latter had been injured when Allied aircraft shot up his staff car.

provide reinforcements. In case the Germans had agents among the French population, deception movements really did take place: 1,200 men with special signs on their vehicles and uniforms made themselves look and sound like an armoured division and three infantry divisions (50,000 men in all) pulling out of the line and heading westwards. In reality, however, American reinforcements were building up all along the southern flank of the salient, at a time when the deception plan of increasing American weakness was persuading Hitler to pour more men into what was soon to become a trap.

Only on 11 August did Hitler realize the danger; he issued orders for a retreat, but by that time no less than eight German divisions were trapped in what came to be known as the Falaise pocket. All but a fraction were killed or taken prisoner. When the Normandy campaign was over, the Allies faced the Germans across the Seine and Paris was about to be liberated.

'In the end' as Chester Wilmot says in his book *The Struggle for Europe*, 'by sending the best part of the 15th Army west at the thirteenth hour [Hitler] had left himself with no organized force to guard the approaches to the Ruhr. In consequence, as Eisenhower says "the enemy was momentarily helpless to present any continuous front against our advances".' Throughout the knife-edge suspense of the whole invasion and the long and difficult battle of the build-up and the breakout, Hitler's best divisions had been sitting waiting only a day or two's march away. There they stayed, waiting for an attack which never came, until the only contribution they were able to make was to arrive too late and to contribute a share of the enormous German casualties.

CHAPTER TEN

BRUSH-FIRE WAR

As THE Second World War drew to its close, its dying embers were to set the torch to a different type of conflict altogether—the local guerrilla wars which came to be known as brush-fire wars, from the lack of warning with which they took flame, the fierceness with which they burned and the near-impossibility of putting them out. They arose in many different forms, from a struggle for power or independence in once-peaceful colonies to full-scale civil wars. Some lasted for weeks or months; others smouldered on for years.

These local wars occupy a fraction of the numbers involved in some of the campaigns of either of the two world wars. But the outcome of many a local conflagration can be just as critical for those involved, to say nothing of the consequences for the rest of the world in an age where confrontation between power blocs is maintained by an uneasy nuclear balance. And for all the advances in weaponry and tactics, in communication and fire-power, the value of psychology—the shrewd blow against morale or the clever deception to catch the opposition off balance—has grown rather than declined.

In one respect the type of psychological operations has changed along with the scale and the theatre of operations. The huge setpiece deception operations of the Second World War are not possible or practical in most of these campaigns—nor are the adversaries or the objectives so clearly defined, as very often the whole battle takes place against the background of a civilian population whose attitudes and support can be crucial in deciding its final outcome.

This was especially true in the prototype of all the brush-fire wars, the Malayan emergency which broke out originally in 1948. This was, in a sense, a complex combination of most of the ingredients which have appeared in all kinds of similar trouble spots ever since. The population was a mixture of two distinct races, Malays and Chinese, and the jungle which covered most of the country made it ideal for guerrilla operations. It had been one of the first territories to be over-run by the Japanese, and during the occupation resistance had only been kept alive under the banner of the so-called Malayan People's Anti-Japanese Army. Despite its name, the MPAJA was a disciplined force of more than ten thousand Chinese guerrillas, led by Chin Peng, secretary of the Malayan Communist Party. Malaya was rich in rubber and tin and, in the vacuum of the Japanese collapse, seemed ripe for take-over. Only the speedy reoccupation of the country and the setting-up of a new British administration thwarted the Communist plans.

So the classic guerrilla struggle began. Five thousand ex-MPAJA Chinese guerrillas, formed into ten regiments all over the Malayan peninsula, returned to the jungle. With them went their weapons, their experience, their organization, the supplies they had captured from the Japanese and the widespread support they still enjoyed from the Chinese community, if not from the Malay majority. By 1948, British withdrawal from India, Pakistan and Burma had shown what could be achieved by firm action, and orders went out to begin a campaign to force the British to leave Malaya in Communist hands.

From the first, however, this war was primarily a psychological war. The guerrillas had to convince the Malays and Chinese that they were going to win in the end, and that the ordinary people had much to gain if they did. The British, on the other hand, had to convince law-abiding citizens, who had no wish to support the terrorists, that they would be properly protected. In the same way, the guerrillas themselves needed hope, encouragement and occasionally compulsion from their own commanders to keep them fighting in lonely and often dangerous situations. If they, too, could be convinced that the security forces would be the winning side, the desire to give up and desert could be a powerful weapon indeed.

The biggest problem, from the security forces' point of view, was the jungle itself. Troops hunting down the terrorists literally had to cut their way through the dense undergrowth to move at all, and do it quietly to avoid broadcasting their presence to the people they were hunting down. Every evening there were soaking, chilling showers, and the darkness was total at night, which was the only time allowed for preparing food and snatching a few hours rest. Wireless contact was impossible without raising an aerial through the treetop canopy, which was so dense as to turn even broad daylight into dim twilight at noon. There were beasts and bugs aplenty, some of them lethal and unpleasant, almost all of them frightening to men unused to the ways of the jungle. And somewhere in the inscrutable gloom were men who were cunning in the arts of concealment, with long experience of jungle fighting. It was a prospect to daunt the spirits of even the most seasoned troops, and morale was not helped by a series of deadly and successful terrorist ambushes.

But the security forces had to take on their elusive and resourceful enemy on his home ground. Led by civilian and military veterans of jungle fighting against the Japanese, the soldiers began a long series of massive security sweeps through the jungle of Johore, near the vital approaches to Singapore. Several guerrilla camps were found and destroyed, but surprise clashes often led to many terrorists escaping, thanks to the lack of experience of the soldiers in these unusual conditions.

This, however, was not the point. The real object of such operations, whether in the Malayan jungle or in any other theatre of guerrilla fighting, was psychological. Even if ambushes failed to kill many of the enemy, the result was that the guerrillas could no longer feel secure in the depths of the jungle. The troops, despite the hideous discomfort of having to live in absolute silence for days on end, in more or less constant gloom, amid rain, heat, leeches and mosquitoes, could at least enjoy the consolation that they were beginning to hit back at the enemy. Following the initial wave of murders of Chinese village headmen and mine managers, a classic guerrilla measure intended to break civilian morale, the local population could draw some comfort from the fact that someone was hitting back on their behalf.

At the same time civilian morale was buttressed by stationing companies of soldiers in the centre of groups of villages, so that any incident in any of these hamlets would bring troops on the scene in half an hour or so. Once again, the benefits were chiefly psychological; even though the terrorists could usually make their escape before the soldiers arrived, at least the latter had a starting point from which to begin the ensuing hunt, and the villagers themselves felt that help was never far away.

The vital point of this new kind of warfare is that the enemy is inextricably mixed up with a civilian population which has to be protected, and encouraged to take its part in the defeat of the guerrillas. Effective military action can only be undertaken when the guerrillas have been isolated from the population, or that part of it whose interests they claim to protect. So, in Malaya, the first steps which helped pave the way to eventual victory were taken by the civilian administration: the local population was brought directly into the struggle by the recruiting of 30,000 volunteer Malay special constables. Not only did this bring in the Malays squarely on the side of the authorities, it also released more troops for offensive operations against the guerrillas.

Other effective steps were the requirement that all citizens above the age of twelve should carry an identity card, bearing their photograph and finger-prints. At the same time, the Government took powers to deport unwanted non-citizens, and to arrest individuals—or even whole villages—if there was good reason to suppose they were aiding and abetting the guerrillas in any way. Finally, the death penalty was introduced for carrying arms or ammunition.

Of course measures like these involve a calculated risk: the guerrillas could make psychological capital out of such restrictions and punishments. But they realized only too well the implications of the tighter control over the population which the new measures were intended to effect, and they fought them by direct and violent methods. They threatened everyone connected with the new identity card programme, from the local officials who issued the cards to the photographers who took the pictures; and, to back their message up, they murdered anyone they could reach who had any involvement with the programme. They destroyed stocks of cards in raids on offices, villages or vehicles. Next they spread the idea that all these new measures were the preludes to Government requisitioning of property, the introduction of new taxes and the beginning of starvation rations for the population. But the authorities stuck to their guns and pushed the programme through, regardless of opposition, and the fury of the terrorists was ample proof of its potential effectiveness.

One of the guerrillas' strongest weapons, in Malaya at least, was the road ambush. Since most troops must always move by road, there was no shortage of targets. And since the jungle bordered the roadside, there was no shortage of concealment. If a convoy was too large or too strongly protected, the ambush party could let it go by, without fear of discovery. When a tempting enough target appeared, they could strike with total surprise. In these circumstances even the most careful planning and the strictest security could not prevent a series of bloody terrorist victories.

There was only one defence, to increase the escorts of all road convoys to the point where they became immune to attack. But this took more troops away from the villages, and straight away the terrorists switched their attacks to isolated police stations, lonely settlements and civilian vehicles. As a psychological blow, this was deadly; for every successful guerrilla attack on a neighbouring village far outweighed in its effect on morale any number of guerrillas killed or captured by army patrols in the jungle.

Something had to be done, urgently, to hit the terrorists at their weakest point, to disrupt their contacts with the civilian population, on whom they depended for support and supply. This was the basis of the Briggs Plan, named after its architect, Sir Harold Briggs, who took over as Director of Operations on 5 April, 1950. This hit at the terrorists by turning the villagers from frightened bystanders into allies, willing or unwilling, of the security forces. The first step was to drive a wedge between the guerrillas and the people closest to them, the 'squatters' or shifting population of mainly Chinese traders, miners and plantation workers to be found living in shacks on the outskirts of almost every village. The only way of isolating the guerrillas was to isolate the squatters themselves in new settlements where they could be guarded and protected, and be freed from terrorist threats.

Building the new villages was a Herculean task in itself. Within two years more than 400 new settlements were built from scratch. To fill them was another matter altogether. As a village was completed, the army would swoop at first light on a particular squatter settlement. The population would be loaded into trucks, often at gunpoint, since rehousing was at first unpopular, and the old huts destroyed to deprive the terrorists of the shelter they offered. The new settlements then had to be protected by trenches and barbed wire and floodlights, and guarded by sentries before the most difficult part of the whole programme could begin. This was the psychological task of persuading the angry and bewildered squatter communities to turn from being allies, however frightened and unwilling, of the guerrillas into aiding the authorities and becoming responsible for their own defences. It was a long and often dispiriting task, but when it worked the rewards were limitless. Tip-offs from informers led to successful ambushes. Troops were told where terrorists would be assembling and would lie in wait in village huts or in civilian vehicles for the guerrillas to appear. Each successful blow was out of proportion to the number of guerrillas killed. The obvious fact that the soldiers were waiting because they knew the guerrilla plans was not lost on the terrorists.

The second part of the Briggs Plan was to restrict food supplies to the terrorists and literally starve them into submission. This involved closing off villages by wire fences, stopping and searching all the inhabitants and workers as they went in and out through the gates, escorting and checking all food convoys, and inspecting and checking all food stores and shops. It was another necessary but unpopular measure. In the long term it could cripple the guerrillas, but in the short term it demanded huge quantities of men. Inevitably, there were less troops and police available for patrols and security duties, and this gave the guerrillas their best opportunity yet. They made good use of it.

June, 1950, saw the guerrillas' biggest offensive. The police were the main target, although rubber-plantation workers were shot and beaten up and villages had crops burned and huts strafed with machine-guns by marauding squads.

Incidents like these struck severely at morale, obscuring the fact that guerrilla deaths were climbing too, as the Briggs Plan and its effects forced them more and more into open warfare, where the security forces could hit back with all their firepower. But the strain was literally killing. General Briggs was forced to retire in November, 1951, and died within twelve months from ill-health and overwork. But the Plan was beginning to work at last; even the squatters were being won over by humane treatment, and the guarantee of their own plot of land of one-sixth of an acre, their own home, pay for five months to cover the costs of their resettlement and help them to establish a new life, and above all protection against the terrorists. One squatter village was so grateful to the troops that they applied for permission to name their village after them—so the settlement of Kampong Coldstream was born.

Food denial, too, was working well. Rice was issued already cooked, as in this form it would keep edible for only two or three days and therefore be useless for terrorist supplies. Convoys were guarded and allowed to stop only at recognized halts under close supervision. Close records were kept of every scrap of food, and in time the restrictions became tight enough to defeat the terrorists' own threats and extortion policy among the fraction of the population they could still reach and dominate.

The Briggs Plan had taken the first steps in carrying the war to the guerrillas. But the process was pushed much further by the appointment of General Sir Gerald Templer as High Commissioner. Templer's methods involved fighting the guerrillas by using their own tactics against them. He emphasized his military background by wearing uniform all the time; he went out to meet the people to an extent previously unheard of. He visited headmen and villages, policemen and soldiers, listening to their problems with sympathy, encouraging them, praising them, and, when necessary, flaying them with criticism. At the same time he backed up his personal example by a new impetus in the campaign to involve the people themselves to a greater extent in the direct battle with the terrorists. He raised more Malay battalions; he raised a Federal Regiment to take in non-Malay citizens, he increased the police and—in a typically audacious gesture of trust which was amply rewarded—he not only raised the Malayan Home Guard, but issued it with its own weapons, rifles and machine guns which were more precious than rubies to the terrorists, but which alone could give the new Home Guard any credibility among its own ranks and the civilian population it was meant to protect.

The other half of Templer's psychological prescription was to strike as hard as possible at the terrorists' own refuge, deep in the jungles. Bombers dropped tons of high explosive on known hideouts, while other aircraft, equipped with high-powered loudspeakers, hammered home the propaganda, with the same theme that guerrillas who gave up and surrendered to the security forces of their own free will would be well treated and paid well for any useful information they could give. Slowly at first the guerrilla deserters began to emerge from the jungle. Once the initial step was taken, the former terrorists seemed surprisingly eager to lead the troops to their former comrades. Often they stayed with the units to whom they had surrendered. When the Green Howards left Tampin at the end of their tour of duty, their ex-guerrillas went with them to their new assignment, and this was by no means unusual.

Those guerrillas who did surrender were used to tempt their one-time comrades in arms to follow their example. Communist propaganda told the guerrillas that anyone trying to surrender would be shot by the British. This was countered by leaflets dropped in millions over the jungle, showing pictures of guerrillas who *had* surrendered looking fit, well-fed, well-dressed and provided with status symbols like brand-new bicycles. Eventually enough one-time guerrillas had changed sides to form a unit made up entirely of such men. Apart from its purely operational value—men with the same training and experience and a thorough knowledge of guerrilla methods—the psychological value of such an obvious large-scale transfer of loyalties was enormous. To see one-time colleagues changing an existence which was lonely, wearisome and dangerous for an obviously secure and well-paid role on the opposing side must have persuaded many still in the jungle either to follow suit, or at the very least give up the struggle and return to normal life. In the end, twelve platoons of Special Operational Volunteers, as these units were called, were raised and armed.

At intervals leaflets and loudspeaker broadcasts from aircraft would

12

announce that the security forces would withdraw from a particular area for a short period, and then tell any guerrillas who might be listening exactly how to go about surrendering. But to make the invitation real, full pressure had to be kept up for the rest of the time, and guerrillas were becoming harder to find.

Some scout units went to great lengths to mislead the enemy. Many stayed in the jungle for weeks on end, with fresh supplies being brought to them by porters, while noisy airdrops were made some distance away to confuse the guerrillas as to the whereabouts of the troops. Ambushes frequently meant lying absolutely still for hours on end, a sore trial in jungle conditions. But by 1955, with independence fast approaching, it was decided to offer the guerrillas an amnesty. Troops had to call on the enemy to surrender before opening fire, and the pressure on the remaining terrorists was relaxed.

It nearly proved fatal. This last-minute respite allowed guerrillas all over Malaya to go back on to the offensive. Attacks were made on camps and plantations, convoys were ambushed and arms, ammunition and food of inestimable value to the terrorists was stolen. All the same, the offer of amnesty for those who gave themselves up voluntarily was maintained. But from the beginning of December the troops went back into the attack as before. At the end of the month, talks which had been arranged between Chin Peng and the Malayan Chief Minister, Tunku Abdul Rahman, broke down in total disagreement, and the amnesty was ended; but the British Government counter-attacked with a promise of full independence for Malaya by the end of August, 1957. This cut the ground from under the terrorists, and civilian support, not to mention their own morale, waned still further.

As the country returned, stage by stage, to a more settled life, the Government was able to go over to the offensive in an entirely different way. By declaring a particular area 'white', the authorities were able to boost local morale by lifting the emergency regulations and restrictions, on condition that the population, through its own efforts and those of its own Home Guard, kept the terrorists away. Malacca was classified 'white' on 3 September, 1953. By the time the country gained full independence on 3 August, 1957, the white areas stretched in a belt right across the peninsula, and the granting of full self-government made nonsense of the guerrillas' original promises to liberate Malaya from its colonial overlords.

Within two months of independence, an offer of a substantial reward for any information leading to the capture or surrender of terrorists was accepted by the terrorists' own leader in South Perak on the west coast. He negotiated the surrender of every man in his area, for 2,000 Malayan dollars apiece. And bring them in he did, in a steady procession over the course of the following six months.

The head of the terrorists in southern Malaya followed his example in the following spring. He brought with him 160 of his men, and by the end of the summer of 1958 most of the southern part of the country was peaceful at last. By early 1960 Intelligence calculated that only a remnant of some 500 guerrillas was left, and that nearly all of these had retreated to the remote jungle near the Thai border, where there was a chance of escape and where they could do little damage. Those who surrendered were now often ill and half starved, and as resistance finally died, the State of Emergency was formally ended on 31 July, 1960. Courage, psychology and dogged perseverance had finally triumphed, winning a notable victory against a resourceful and determined foe who began the fighting with most of the natural advantages on his side.

In essentials the Malayan campaign was to provide the model for most of the conflagrations in which British forces have been involved. In Kenya, where the Mau Mau guerrillas began the familiar progression of burning, wounding and killing in 1952, the scale of the problem never approached that of the Malayan emergency. But although the physical tactics were modified to deal with a different enemy in a different theatre of war, the psychological approach followed the same basic precepts.

The Mau Mau outbreak was confined to a single tribe, the Kikuyu. But within the tribe the guerrilla organization exerted a powerful mental grip for two reasons: there was already much bitterness over the shortage of good farming land, which the terrorists could use as a cause to secure the support of their own people, and the psychological make-up of the tribe was such that oaths and ceremonies carried a great deal of weight. By imposing their own lurid ritual on all who joined or supported their bands, the Mau Mau forged bonds which took a great deal of breaking, however much those who had taken their oaths might want to change their allegiance. Innate superstition was backed by the threats of horrible deaths for oath-breakers, and the whole apparatus reinforced by ritual murders of those suspected of lukewarm allegiance or outright disloyalty.

At the beginning of the campaign the Mau Mau was fairly secure in its support among its own tribe—however grudging and reluctant in the case of the majority of the population. On the other hand, tribal loyalties being so close-knit, the Mau Mau troubles were sharply localized. People from other parts of Kenya were relatively untouched, and since few of the Kikuyu enlisted in the armed forces, the two battalions of the King's African Rifles which were recruited in Kenya were free of the threat of subversion or mutiny. Together with six more battalions sent in from elsewhere in East Africa, they were to play a vital role in the campaign.

The storm broke on the first day of 1953 with the murder of two white farmers. From then on the attacks and the killings grew apace, and the hold of the Mau Mau bands over their tribe grew steadily tighter. Reinforcements were sent in the form of two more battalions of infantry, and the security forces prepared to take the offensive, but before they arrived the Mau Mau carried its campaign of terror against its own people a step too far. On the same night, 26 March, 1953, as one gang carried out a successful attack on a defended police post to seize arms and release prisoners, another attacked the Kikuyu village of Lari, twenty-five miles north-west of the capital. Eighty-four people, including women, children and babies were literally hacked to death, and the thirty-one survivors were left mutilated and close to death. It was an attack which in the end was to be fatal to the Mau Mau movement, for its effects were to create an anger which in many cases overcame the deep fear of the consequences of oath-breaking or of direct action against the gangs.

Before long progress was being made along similar lines to the process which was working well in Malaya. Identity cards were issued to all male Kikuyu, and the Kikuyu Home Guard was raised. Recruiting at first was slow, but the horror within the tribe caused by the Lari massacres saw its numbers double to a total of 20,000 within weeks. To begin with these brave volunteers, who were putting their faith in law and order and duty to their fellow men above their very strong tribal loyalty, were armed only with spears and pangas. In time, when their reliability was well proven and when weapons could be spared, they were properly armed, and they became an immensely valuable aid to the authorities in keeping the Kikuyu area policed.

Another consequence of the Lari massacre was that information began to flow in to the authorities, and this—as in Malaya—was of incalculable value in planning raids against the Mau Mau. The thick forests of the Aberdare Mountains and the foothills of Mount Kenya, happy hunting grounds for the guerrillas, were declared Prohibited Areas. Here anyone not a member of the Security Forces could be shot on sight.

Roads were cut through the forest and patrol camps built as far as four miles into the woodland. This made surprising the guerrillas much easier, and the psychological attack was intensified by a heavy programme of bombing by the RAF. Although the forest covered a huge area, the absence of civilians allowed attacks wherever Mau Mau bands were thought to be operating, and later interrogation of prisoners showed that the bombing had had a great demoralizing effect, forcing them to keep on the move to avoid attacks.

As each area was swept, it was declared 'white' on the Malayan basis, and its own inhabitants made responsible for keeping it free of guerrillas. Isolated and vulnerable Kikuyu settlements were removed to new villages with proper defences. And as the gangs were pushed into the forests, they were hedged in by a deep ditch, backed by barbed wire entanglements and booby traps, and guarded every half mile by police posts. At the same time, as more and more troops were freed from duty in the settled areas, they were able to cut off all food supplies to the gangs still at large.

But starving them, and forcing them into retreat was only part of the battle. Capturing them was increasingly difficult as their numbers dwindled. In 1955 the security forces began, first, to use surrendered Mau Mau men to lead them to their former comrades, and then, to form units of former Mau Mau men to act as pseudo-gangs and lay ambushes of their own. At first they were led by Europeans, but from the middle of 1955 onwards these new allies were sent into the forest on their own. Although there was a risk that they might desert en masse with the weapons given them by the authorities, they succeeded in smashing the real gangs which still survived.

After Malaya and Kenya each new campaign was to bring its own individual problems, but there were still important common factors. One of the most vital was the need to win over the local civilian population to the side of the authorities. In Borneo, in 1962, when Indonesian guerrillas began crossing the borders into Brunei and Sarawak in support of President Sukarno's 'confrontation' campaign against their future amalgamation into the independent Federation of Malaysia, no effort was spared to boost the morale of the villagers. Extra medical aid, by helicopter-borne army doctors, was backed up by new agricultural and economic help plus purely morale-boosting events like the visits of regimental bands to remote frontier villages. At the same time many offensive operations had to be modified and reduced in scale because of the unacceptable moral cost of a single civilian killed by accident. Because of the long frontier and the ease of mounting a surprise attack the guerrillas were usually successful with their initial blows. But, as the army were seen to be quick in their retaliation, sealing the escape routes and usually catching their enemies, morale remained fairly high. Local villagers were trained in self-defence; troops were established in bases deep inside the jungle, and patrols were mounted whenever intruders were reported, sealing off their escape routes and laying ambushes when they least expected it, when their vigilance was relaxed on the way back to their own territory. These tactics became even more effective when the Indonesians began parachuting their men into Malaysian territory, and very often the guerrillas found the native population not only unexpectedly hostile but well capable of defending themselves, both as a result of the Forces' backing and training.

By the middle of 1965 units were striking deep into Indonesian territory, hitting the guerrillas where hitherto they had been safe. The psychological impact of a hard-hitting attack when they were busy making plans for their own operations, or when they were resting and refitting after an escape from earlier ambushes on the other side of the border, was redoubled. Other tactics included landing 105 mm guns by helicopter deep in the jungle for attacking enemy camps. When the shells began to fall the guerrillas vanished into the jungle, only to find infantry posted on all the trails in ambush positions. Soon this became so well known that the opening rounds of the artillery bombardment produced total panic and despondency among the defenders.

By September, 1966, the situation was stable enough for the British and Gurkha troops to leave North Borneo to the Malaysian Army. But the reason for the triumph had been the psychological victory which had made the military victory possible. For all the courage and endurance of the forces, the first priority had been, in Templer's words, 'to win the hearts and minds of the people', just as he had applied them in Malaya. From the beginning, great care had been taken to encourage the local population to see the troops as their friends and allies, to realize that they had nothing to fear from Indonesian threats, and that they could and would be defended, provided they played their own part. The support of the villagers enabled the security forces to take the offensive against a determined and at times extremely tough and resolute enemy, and gave them the time and the backing they needed to complete a long and dispiriting task. Like Malaya, it had been a battle of attrition as much as had Verdun or the Somme, except that now the decisive factor was psychological rather than material attrition. The Indonesians turned against Sukarno's policies not because they were physically beaten, but because psychologically they felt they could no longer win.

Other campaigns provided greater challenges, both materially and in terms of the war of morale. Usually this was because the local population, or a sizeable part of it, were actively hostile from the beginning, as was the case in post-war Palestine.

The Jewish campaign against the British administration began on the last day of October, 1945, with explosions cutting railway lines, sinking ships and damaging oil refineries. Within days the security forces found themselves up against the most daunting weapon of all—the totally unyielding unity and hostility of almost the entire population, which made searches and interrogations almost pointless. This was the ideal environment for the guerrilla, safe in the midst of a united and sympathetic community. Riots broke out in Jerusalem and Tel Aviv, but here psychological victory went to the soldiers. They came to relieve the over-extended police with bayonets fixed and notices announcing 'Disperse or we fire' in three languages. The crowds in Jerusalem soon dispersed, but those in Tel Aviv returned to the attack on seeing that the soldiers were staying where they were and not pursuing them. Only when sections of troops, in box formation, started clearing the streets one by one did violence break out again. Shots were fired and one civilian was killed before the crowds dispersed. The battalion involved was soon reinforced by four more and, after a tight curfew for four nights, order was finally restored.

But other tactics were more difficult to counter. Whenever the police tried to raid villages to take suspects for questioning, bitter resistance forced the troops to go in to support them. Sometimes they would be charged by furious mobs; at others they would be confronted by sullen and passive resistance. And the raids were increasing. RAF aircraft were blown up on their own airfields; armouries and police stations were attacked. Any raid which failed was a tonic to morale, like the raid on the 3rd Hussars' camp at Sarafand on 6 March, 1947, when a Jewish raiding party arrived during the lunch hour in a British army truck, disguised as paratroops. The alarm was raised and the truck tried to escape, but it was later captured along with two of the raiders.

But incidents like these were rare, and progress was slow. Information from within the Jewish community was virtually non-existent. Recruiting local people into a Home Guard to help reduce violence was only successful among the Arabs, with the result that this force became identified with one side rather than with the community as a whole.

If any terrorists were captured, the usual retaliation was to capture army personnel as hostages for their lives, a tactic which was aided by the army trying, for good psychological reasons, to maintain a low profile by presenting as relaxed and as unwarlike an appearance as possible when off duty, to avoid provoking even more hostility among the population — many of whom had suffered so much at the hands of armed soldiers in Europe so very recently. Other tactics which were to work well in Malaya and elsewhere misfired badly. Snap searches, backed by the most meticulous preparation, failed to capture the men who were directing the terror campaign — and a conciliation campaign, which tried to ease the tension and split the opposition by the release of the more moderate Haganah prisoners, resulted in the more extreme Irgun taking the offensive even more aggressively. A new wave of bombing began in November, 1946, aimed at the railways, and the Arab engine crews in particular. This was a highly effective campaign. As disaster followed disaster, the train crews refused to move save in daylight under heavy military escort, which used up large bodies of troops in simply keeping the communications working. At the same time, cunning booby traps were taking a terrible toll of the officers and men detailed to defuse bombs. The forces were living more and more under seige conditions, in barbed-wire-protected camps and always under a dampening rain of hatred from the population. One solution was to move units around as much as possible. This helped morale, but it wasted much valuable experience built up by any unit which had had time to acquire knowledge of its own particular area. The problem was intensified early in 1947, by the moving of the civilian administration into compounds under police and military guard, following more bombings and kidnappings of hostages to be used as bargaining counters for the lives of terrorists under sentence of death.

During the course of 1947 the attacks grew ever fiercer and a new psychological tactic was used against the British forces. Whenever the guerrillas mounted a large-scale attack against the Arabs, they made ample use of their stocks of captured British trucks and uniforms. The Arabs were not fooled, knowing there was no reason for the British to attack them. But several genuine British columns were set upon by Arabs who were convinced they were disguised Jewish columns. Other Arab forces were massing on Palestine's borders. The first of the invading columns was turned back by psychological pressure from deliberate near-misses from mortars operated by paratroops and from dummy attacks from low-level RAF fighters. Other Arab columns were willing enough to turn back when firmly requested to do so — provided the army would oblige them with a convincing bombardment, so they could claim to have been overwhelmed by vastly superior forces.

Right up to the final evacuation, the Irgun kept up its private war against the British. Camps were shot up, prisoners taken and executed, patrols were ambushed and on one occasion two wounded soldiers were followed to hospital and there shot by a gang of terrorists who followed them into the wards. The army, meanwhile, was trying to keep the country together with parades, flag marches and visits to villages and settlements. They received hearty welcomes from the Arabs, contrasted with sullen distrust from the Jewish population, and it was hardly surprising that many of the men became less than impartial. Yet this was to lead to more trouble. Any preference shown to the Arabs would only result in still more alienation from the Jews, and only by working with both communities could any real solution possibly be achieved.

Cyprus, too, was bristling with difficulties. As with Palestine, the population was divided into two different national groups, which became increasingly suspicious of one another. The majority of the population was firmly against violence, so the Greek guerrillas had less direct support than they might have wished, but on the other hand community loyalty and drastic reprisals against informers prevented them from being as helpful to the authorities as had been the case in Malaya.

There were similarities, though. Where the bands led by Chin Peng had lived in the jungle, Grivas's men hid in the wilderness of the Troodos Mountains. Where the Mau Mau forced their supporters to swear oaths in blood, EOKA'S men swore in the name of the Holy Trinity, a powerful commitment among a still very religious people. There were also differences. EOKA began their full-scale campaign with mobile, hit-and-run raids on the classic pattern, but the security forces reacted quickly by carrying out a series of sweeps. So, from 1956 onwards, Grivas changed his tactics to raids and assassinations within the towns. The population was kept firmly in hand by a ruthless policy of intimidation. Those who were held to have broken their oath of allegiance were summarily shot. And carrying the fight to the towns gave EOKA another advantage – any moves against them would be bound to affect the ordinary population; thus British actions could be made into useful propaganda of colonialist oppression of innocent people.

This was a much more difficult proposition for the security forces to deal with. The sweep which had driven EOKA into the towns – Operation Pepperpot – had been a mixed success. A dozen guerrillas had been captured and the power of the mountain bands had been seriously damaged, but Grivas's escape was giving rise to a legend among the people that he was uncatchable. Public opinion seemed to be turning more and more in favour of EOKA, although many government officers still felt this was due to powerful intimidation behind the scenes, and carefully organized 'spontaneous' demonstrations of sympathy.

But in time there were grounds for hope. By 1956 Athens Radio's daily outpourings of anti-British propaganda had been silenced by heavy jamming, and the security forces had almost cut off the supplies of smuggled arms from outside the island. Enough disgruntled ex-EOKA men and enough determined opponents of Grivas had been found to form counter-gangs which carried the war to the enemy in the same way as their predecessors had done in Malaya and Kenya. Under all this psychological and physical pressure, the tempo of terrorist operations began to slacken at last.

In December, 1957, Sir Hugh Foot was appointed Governor, and he began winning over the Greek-Cypriot population, as Templer had done with the Malays. His frankness and sincere desire for a settlement may well have won a great deal of sympathy among the Greek community, but his words were anathema to the Turks who were now convinced they would be sold out as part of the bargain to be struck under the threats of renewed violence by EOKA.

In March, 1958, the truce broke—EOKA blaming the army and the British blaming EOKA. The psychological campaign was stepped up by the terrorists in a last bid to force a settlement on their own terms: the population was ordered to boycott British goods, British property was blown up, pupils were taken away from British schools, and the shooting began again. Every time negotiations began on a new set of proposals for the island's future, another truce was announced. Everyone knew that once the talks broke down the shooting would begin again. In June, 1958, the Turks started a war of their own against the Greeks, and once again the British were caught in the middle, trying to keep the communities apart.

By November, 1958, the population on both sides was growing increasingly weary of the struggle and the never-ending negotiations. Psychologically, this was the moment for compromise—and just before Christmas, 1958, the Government of Cyprus reprieved four EOKA men under sentence of death. Grivas responded with another cease-fire, provided the army also called off the fight. In fact the troops went on searching but the terror campaign did not recommence. The Greek Government, worried by the violence of the Turkish reaction, dropped its support for *Enosis*, and Grivas's most desperate tactic, the campaign against the Turkish Cypriots, rebounded with a vengeance. After all the heroism and all the terror, all the violence and all the pressure, the final solving of the Cyprus question was left to the Greek and Turkish Governments, who decided on partition. Nobody had gained what they really wanted, although there was something for everyone in the final agreement: the Turks had protection within their own areas, the Greeks had increased links with mainland Greece, and the British had their bases. What had been for generations a peaceful and happy island was now a bitterly divided community. Fighting between the two races broke out again in 1963, and once again British troops had to keep the communities at arms' length until the United Nations could take over. Even this was only a temporary solution. Increasing fear and bitterness among the Turkish community, and warlike threats from mainland Greece culminating in large numbers of troops being sent to the island, resulted, in 1974, in a full-scale Turkish invasion and the seizure of the northern part of the island as Turkish territory. After twenty years, the problem of Cyprus still defies solution. It had been perhaps the most difficult campaign yet—but, both militarily and psychologically, the security forces were unbeaten.

Palestine and Cyprus both presented the security forces with totally new psychological problems, of a kind soon to become depressingly familiar nearer home, with the emergence of a new and deadly enemy — the so-called urban guerrilla. Combating this new menace demanded new tactics. In Aden, for example, traffic into the centre of the city had to be channelled through checkpoints where troops could search for arms and explosives. Soon the sentries became specialists in out-thinking the terrorists over places of concealment, checking insides of spare tyres, contents of pots and pans and loads of fruit and vegetables piece by piece to prevent smuggling of supplies. The physical sense of being hemmed in also worked in the soldiers' favour: terrorists caught in checks and snap searches gave information which led to more searches, and the haul of captured arms and ammunition grew apace. The soldiers introduced methods which had worked well in Cyprus. Isolated groups or single men would present themselves as tempting targets, but any attempt at grenade-throwing or ambush soon revealed reinforcements lying in wait. Houses were stealthily fortified as guard posts, and soon the terrorists avoided the obvious targets on principle, expecting ambush parties where none existed. Later the reactions of the troops and police grew even more uncanny. Thanks to observation posts on the slopes overlooking the Crater district, terrorists could often be spotted through powerful telescopes, and police and soldiers radioed to the spot for snatch arrests.

Many of these techniques were to prove invaluable in out-thinking the enemy in the army's latest and most difficult psychological challenge of all: Ulster. Many of the problems facing the forces here are familiar enough, even if they are more severe than before. Once again, the army's major opponent is a force representing only part of the community, so that there is little that can be done to win over that part of the population without alienating the remainder, as happened in Palestine and Cyprus to differing degrees. On the other hand, Ireland's history, and legacy of deep bitterness towards the British Army as an instrument of British policy and an obstacle in the struggle for Irish independence, adds one complicating factor. Another is the inefficient partitioning of the country, which left a large Roman Catholic and Republican-minded minority living under the rule of a Protestant majority inside Ulster, which itself is far outnumbered by the population of Ireland as a whole. The chances of finding the right kind of psychological formula for the army are almost non-existent.

Yet attempts were made to win over both sections of the population as an essential first step to maintaining order. The first soldiers who arrived in 1969 were hailed as saviours by the Catholic population. The police, being drawn mainly from the Protestant community, were regarded as their traditional opponents, but troops from Britain were a new factor, and they were given the benefit of the doubt, at least in the beginning. So again, as in Cyprus, peace lines had to be drawn up, and etched out in barricades and barbed wire, to keep the protagonists apart. The lines then had to be patrolled by troops, which left them highly vulnerable to attacks from either side.

For a while all was well. But the IRA began a very shrewd psychological attack by mounting its initial campaign against the Protestant community rather than the army itself. This, they knew, would force the army into acting more and more in defence of the Protestant community. The more this became obvious, in the emotional and highly polarized logic of Northern Ireland, the more the Catholic community would see the soldiers operating alongside the police, and would come to identify them as another enemy, rather than impartial representatives of law and order. The IRA were then able to pose as the only defenders of their own minority community, a tactic which won them much more support from the civilian community than they might otherwise have had. The IRA developed formidable tactics of its own. Troops on riot duty made excellent targets for snipers: sometimes agents

within the crowd manœuvred the rioters to draw the soldiers into the
fire of carefully planned ambushes. And, just as in Palestine, the
anxiety of the army to keep a low profile left off-duty soldiers highly
vulnerable to attacks designed to sap their morale. The murder of three
Scottish soldiers in a country pub, the luring of another group to a flat
by some girls on the pretext of a party, where IRA killers were lying in
wait, the constant sniping and ambushes, all were calculated to make
the troops angry and dispirited.

Psychologically too, the IRA were to prove formidable opponents.
Allegations of torture were made by prisoners after interrogation;
occupants of houses where surprise searches had been made com-
plained of intimidation and threats from troops, even when their homes
had concealed stocks of weapons and ammunition. Every act by the
troops was turned into propaganda material—not only among many of
their own people, for whom the army was the eternal and natural foe,
but in world opinion also. The so-called 'Bloody Sunday' incident,
when the army struck back at concealed snipers cleverly capitalizing on
rioting crowds, was trumpeted to the world as a deliberate massacre of
an innocent population—and, since mud sticks, the result was a boost in
local and international support for the IRA.

But the army too has had its real successes. When worsening relations
between the communities resulted in the Catholic communities denying
access to many of the areas they felt most threatened, the IRA took
over their defence, announcing them as 'no-go' areas where the writ
of the IRA itself was law. This too was excellent propaganda—here they
were the actual as well as the psychological defenders of their own
community against outside attack. But this was a heavy risk for any
guerrilla movement, committing itself to the defence of specific
objectives. Only the conviction that the army would not risk a full-scale
attack against defended urban areas, for fear of heavy casualties and
disastrous propaganda, persuaded them to take that risk.

Here, however, was one occasion where the army more than out-
guessed its opponents. Just as in Kenya, Cyprus or Palestine, the
operation against the no-go areas was planned in conditions of absolute
secrecy. Disguised officers reconnoitred as closely as they could the
objectives their men would be called upon to capture, and routine
movements were altered to hide any signs of the preparations for the
assault. In case any information did leak out, rumours were allowed to
hint at some kind of operation which was much further off than the real
thing. Preparations were made at night, and at first light the army
moved in with a speed and practised ease which spoke well of all the

experience gained so painfully in similar situations abroad. Operation Motorman, as it had been christened, was a great success. With scarcely any real resistance, the army was master of the areas the IRA had sworn to defend to the death. The propaganda value, not to mention the psychological boost to the troops themselves, was incalculable.

Other successes too came the army's way. One advantage of the Irish situation is that, with many of the army's own men coming from the very communities among whom they operate, the chances for undercover operations are limitless. Occasionally things go wrong — and shooting incidents involving soldiers dressed as milkmen or house-to-house delivery men must imply a great deal of covert intelligence-gathering.

But in psychological terms the war in Northern Ireland is one of attrition. In many ways the army must be on the defensive in this side of its operations, since the winning over of the civilian population — the first requisite for success — can only be a long-term objective. In direct military terms, the army has kept up crushing pressure on its adversaries, which has forced them into making mistakes in psychological tactics. As more and more of their men are killed or captured, the IRA must come to depend more and more on the civil community for support and concealment. On the other hand, once people begin to withdraw that support in disgust at the continuing violence, the IRA can only do what Grivas did in similar circumstances — escalate the terror campaign against those of their own people who do not toe the line. But while this may preserve their security in the short term for a little longer, there is little doubt that it erodes their support in the long-term.

As this book goes to press, the struggle in Ulster continues unabated. Yet there are perhaps signs that the psychological battle is beginning to go the army's way, even if many of the factors which influence this are outside its direct control. The rise of the inter-denominational Women's Peace Movement, uniting both communities in a denunciation of violence and those who practise it, has put the IRA in an extremely difficult position. It can argue and threaten, but any direct action against such a group, however threatened it may feel, can only lose it more support than it has already lost through increasingly horrific bombings and shootings.

Other outside factors are adding to this effect. Traditional sources of funds and weapons, like the Irish communities in America, are beginning to re-think their attitudes following the Eire Government's deeper commitment to fighting terrorism. This, too, robs the IRA of its once-safe bases on the other side of the border. As support wanes, this creates the hope of other tactics which helped win the battles in other campaigns—principal among them the faster flow of accurate information, the lifeblood of anti-terrorist operations anywhere. Since the campaign is still continuing, detailed information on why and how these operations are mounted, and what tactics and deceptions have been evolved, cannot be quoted without losing much of their value, but the presence of SAS units in areas where fanatical Republican support makes it possible for IRA assassination squads to operate more in the open makes some kind of counter-gang operation on the Malayan, Kenyan or Cypriot model a distinct possibility.

The army's priorities are the same as they have always been in this kind of fighting. They must convince the Catholic population that the soldiers are able and willing to protect them against intimidation. Telephone lines for anonymous informers help people, give information to prevent bloodshed, but they are frightened of becoming victims in their turn. The troops' role in defusing bombs and preventing explosions has won a great deal of public sympathy, but their prime objective remains: to go on fighting the IRA so effectively that the terrorists lose their will to win, and above all their confidence that they *can* win in the end. Undercover agents within the internment camps and prisons can whittle away the morale of IRA prisoners, fomenting discontent between those on the inside and those who have taken their place in the fighting outside. Rumours of badly-run operations, illicit affairs and misapplication of funds or weapons can have a deadly effect in a clandestine organization where real information is often delayed or cut off altogether.

One thing is certain: whether the army succeeds or fails in Ulster depends above all on the psychological battle. Calculated risks have to be taken, like the acceptance of truce offers which allow the terrorists time to regroup, or the release of detainees from prison, despite the military and tactical risks. Even if they never succeed in convincing dyed-in-the-wool Republicans that the soldiers are their friends, if the army can only persuade them that the IRA cannot and should not win this battle, then its task will be half over. But the pressure is still great. The IRA continues to try every trick in its considerable repertoire to portray the actions of the troops as aggressive, dangerous and prejudiced against the Catholic community. The wrong action in the heat of the moment, in the thick of a riot or under deadly sniper fire, can undo the good effects of months of patient work.

As full-scale conventional war becomes more unthinkable guerrilla campaigns involving this kind of psychological confrontation will become the pattern for the future. So the techniques and tactics of psychological warfare will become all-important, although, as in any other branch of warfare, the lessons and ideas of the past will merely be starting points.

CHAPTER ELEVEN

COLD-WAR CONFRONTATIONS

SMALL-SCALE GUERRILLA battles may have become the pattern for war in the 'forties, 'fifties, 'sixties and 'seventies – but psychological warfare on a global scale has never quite died out. In fact, it is only because of the continuing balance between the super-power blocs that the conditions which favour guerrilla campaigns exist – and a vital factor in maintaining the continuing balance on which the peace of the world depends is the psychological and deception battle which rages unabated even in the absence of hot-war fighting.

Yet the conditions on which the Cold War was based depended not on either America or Russia or indeed any of the Allies, but on the deceptions and stratagems of Hitler's Germany in the months when the Reich was tumbling to defeat. Although the end may have seemed inevitable to the Allies, once the elaborate but successful D-Day deceptions were over and the *Wehrmacht* had been pushed back to its own frontiers, inside the fevered atmosphere of the higher Nazi councils anything was still possible. New weapons, from rockets to jet fighters, from nerve gas to super-submarines, were under development. All that was needed to reverse the fortunes of war was time. And gaining time could be done by any and every trick to worry, confuse and above all delay the advancing armies now closing in.

Hitler's first attempt to fight back arose directly out of the Allies' success over the D-Day invasion plan. The one vital flaw in attacking through Normandy rather than the Pas de Calais was so obvious that Hitler himself had seen it as a reason why the Allies could not *afford* to land in Normandy – avoiding the Pas de Calais gave the Germans the chance they needed to launch the first of the so-called revenge weapons against London. Starting exactly a week after the Normandy landings, at a rate of one launching every six minutes, the V1 flying-bombs thundered off their ramps, aimed straight at the centre of London.

Yet, fortunately for the Allies, the Germans' own deceptions had been far less effective. Although development of the V1, and of the V2 ballistic rocket which succeeded it, had been shrouded under the strictest secrecy, the Allied intelligence services had been picking at every threadbare patch in the German security blanket until, by the time *Oberst* Max Wachtel's 155th Flak Regiment fired the first operational V1 to be aimed against England on 13 June, 1944, they did at least have a sound idea of the threat confronting them, and how the danger could be lessened.

For the moment the flying bombs seemed to be the more urgent problem, and new defences were hurriedly set up on the approaches to London, from radar-controlled batteries of anti-aircraft guns to squadrons of the RAF's fastest fighters – Spitfire Mark XIVs with the more powerful Rolls-Royce Griffon engine, and Hawker Tempests, the latest interceptors to enter service. Finally, on 15 June, two days after one of the first V1s had killed six people and injured nine when it fell in Bethnal Green, the V1 offensive proper began. On that day alone, 77 V1s crashed down on London, having evaded the guns, the fighters and their own mechanical malfunctions. Hitler's revenge was beginning to get into its stride.

But from the German viewpoint, there was one serious drawback. The aiming of the flying bombs was crude at best, so that only a target the size of London offered any real attraction. And every V1 which left the launching ramp on its way to the capital literally vanished into the unknown. Unlike normal artillery fire, there was no way in which the Germans could observe the effect of their weapons, nor correct their aim to use each one to better effect.

Or perhaps there was. What about the agents who had been so devotedly sending out information on Allied troop movements and invasion plans? Could they not also report on where the flying bombs were landing? Enough reports could be used to produce a pattern which would show where the mean point of impact was centred, and if a suitable shift gave a more damaging spread, then the control settings given to the V1 launching crews could be altered. So Garbo's controller, having warned him under the strictest secrecy to move to the northern side of London as early as the end of 1943, now begged him for every detail of V1 explosions which he and his sub-agents could find.

Already, just three days after the beginning of the bombardment, some hard facts about the V1 were emerging. Although some fell wildly off course, over half were landing within a seven-mile circle, and the mean aiming point was off-centre. Some error in calculations had shifted it from central London to the south-eastern suburbs near

Woolwich. This was just the information the Germans were after: correcting the control settings to shift the impact north and west would be a matter of minutes once they had the information. Then the crowded centre of London, with all the Government offices, the major centres of population, the railway termini and the Thames bridges, would receive the full weight of the bombardment instead of the occasional overshoot.

But if they were given the *wrong* information? If all the reports they were sent referred to hits on the northern side of London, then they could only conclude that the Vis were overshooting their aim. They would have to alter the settings to reduce their range, and the bombs would fall in lightly populated areas, many of them in open country. So a special operations room was set up to plot every Vi incident. From these were selected enough in the right area to create the right impression of overshooting.

After initial resistance in the Cabinet, the plan was put into operation. Already the mean point of impact, when plotted against all the incidents, was shifting to the southwards, just as the deception team of the Crossbow Committee (the committee chaired by Duncan Sandys, which had been set up at Churchill's instructions to evaluate the German V-weapons and devise the best counter to them) had intended. Then came a totally unexpected bombshell, drawing attention to a vital loophole no one had thought about before, which might easily show the Germans exactly what was happening.

It emerged that a civil servant had had a letter from his wife, who had been evacuated to the West Country, in which she commented about the bad time they seemed to be having from the flying bombs in *South* London. The deception team were alerted, and the writer of the letter questioned on how she knew *South* London was suffering so badly. Her reply was that she had noticed that most of the people quoted in the *Daily Telegraph* or *The Times* as having 'died suddenly', a euphemism for those killed by bombing, seemed to have South London addresses. She had collected a number of these and plotted their positions, which gave a mean point of impact almost exactly the same as that selected by the Crossbow team.

If it was as easy as that, why did the Germans not spot the loophole? Copies of English newspapers were easy enough to come by. Hurriedly the censorship regulations were adapted to change the wording of the death notices, and the deception plan went on. But as an extra precaution to protect Garbo, and by implication his whole team, from being too deeply involved in a deception which might later be exposed, his scriptwriters contrived to get him 'arrested', while snooping for signs of bomb damage in the London docks. One of his fictitious sub-agents sent a message to the Germans, and a few days later they were relieved to hear that Garbo had been released. But he now had an excellent excuse for staying well out of the limelight until the danger had receded.

Eventually the V1 threat dwindled. Practice made for better gunnery, and fighter pilots grew more adept at catching the flying bombs and either shooting them down or tipping them over to tumble the gyros and send them spinning harmlessly down over farmland. And as the German armies were forced further back the launching sites themselves were captured. The V1 was a strictly limited-range weapon, which was designed to be fired from prepared sites. Once these were lost the days of the flying bomb were over.

There was still the V2, in the long term an infinitely greater threat, since it could be fired from portable launchers and, once the missile was on its way, there was no defence against it whatsoever. But, as the German forces were pushed further and further back, by early 1945 the last areas under their control from which rockets could reach London had been lost.

By this time, too, Hitler's last military gamble had failed. As the fighting moved into Germany, he had begun to draw up plans for one last great master-stroke to throw the Allied advance out of gear, split their front and drive a wedge through them to capture Antwerp and force them back to the sea. This would gain Germany time to bargain, time to exploit growing dissension between the Western Allies and the Russians, time above all to build more of the miracle weapons which would turn the tables once and for all. But over all the Western Front the Allies had air superiority. How could any preparations for such an offensive take place without the enemy knowing about them as soon as they started?

Hitler had been determined to use the Ardennes as the launching-pad for his new offensive since the middle of September, 1944. It was essential that no soldier belonging to the attacking units brought in for the offensive (whose presence on this part of the front was not even suspected by Allied intelligence) should risk being captured by an Allied patrol. So they only moved into their final positions on the night before the attack. The offensive came as a complete surprise to the Allies, and confusion was made worse by another weapon from the Germans' armoury of deception—Operation Greif. This was master-minded by Colonel Otto Skorzeny, the man responsible for rescuing

Mussolini from his mountain prison in 1943, and later for the kidnapping of Admiral Horthy, the ruler of Hungary, who was preparing to surrender to the Russians in October, 1944.

This time Skorzeny was to recruit a brigade of two thousand German soldiers, either of American descent or who had lived for some time in the United States. They were dressed in American uniforms and equipped with captured American tanks, trucks and jeeps. Once the breakthrough was under way, Skorzeny's false Americans began to spread out ahead of their advancing compatriots, through the bewildered and retreating enemy forces. Sometimes they hit at communications by ambushing staff cars and despatch riders, or by cutting telephone wires; sometimes they spread confusion by misdirecting traffic and giving false messages. They also tried to pinpoint the locations of the vital Allied fuel dumps which the Germans had to capture intact to maintain the momentum of the offensive.

Some of Skorzeny's men went ahead as far as the bridges over the River Meuse. Their job was to prevent them being blown up by the Americans before the Panzers could reach and cross the river. But in fact the Germans in American uniforms were the only ones to reach the Meuse, for the Americans of 1944 were already proving themselves a very different proposition to the French of 1940. Here and there, isolated and outnumbered groups continued to hold out against furious German attacks, each one slowing and impeding the German offensive. Slowly, imperceptibly at first, and then more obviously, the German advance fell behind its timetable.

The Allied intelligence organization had its first stroke of luck on the very morning of the attack. One of the first Germans to be captured by the Americans was carrying several copies of the plans for Operation Greif, an extraordinary mistake in view of the great care taken over all other aspects of security. And although this alerted the Americans to the German deception, it also caused them a lot of worry and a great deal of trouble, since some of the first Skorzeny commandos to be captured mentioned rumours that some of their comrades were aiming for the Allied Headquarters to assassinate General Eisenhower and members of his staff.

Suddenly everyone in American uniform was a potential enemy. Order was followed by counter-order and rumours ran wild. From the front line right back to Paris, officers could only be obeyed by those who knew them. Strangers were confronted with demands to name the baseball team or the capital of their home state, or assailed by other questions designed to check their origins. The confusion became worse, and the whole organization of the American armies faltered for a short time on the edge of chaos.

In the end it was a close-run thing. Battlegroup Peiper, the advance guard of the 6th SS Panzer Army, got within a mile of Stavelot, where there was a petrol dump containing literally millions of gallons, but came to a stop for lack of fuel. And a reconnaissance regiment of the 2nd Panzer Division came within three miles of the Meuse before stopping to refuel. By now, however, time was running out and the bad weather was beginning to lift. As soon as it did, then the terrible power of the Allied air forces returned to tilt the balance for ever against the Germans. From now on every supply lorry which ventured along the roads to the Panzer units was shot up. Finally the troops who had advanced furthest were compelled to abandon their tanks in the snow, undamaged but immobilized for lack of fuel. After a week of hard fighting the retreat began, and within a month the Germans were back where they started, having lost half their men and most of their equipment—double the American losses, and at this stage of the struggle completely irreplaceable.

So the Ardennes offensive, on which so much had depended and for which so much had been hoped, ended in ignominious failure. The Allies had been caught napping by clever preparations, but the result was worse for the Germans than if it had never been embarked upon at all.

Yet there was one last card left to play. And in a way it was fitting that it should be Dr Goebbels, Minister for Propaganda and past master of the arts of propaganda and persuasion, who had to play it. The deception was fraught with irony. The Allies, who had played so deftly on the strings of the Germans' own prejudices and preconceptions with their various plans, not only fell for Goebbels' suggestion hook, line and sinker, but went on developing it and embroidering it of their own volition, every bit as blindly as the Germans had done with the invasion plans.

As the situation worsened for Germany, and losses could no longer be hidden, Goebbels took on the task of heartening the population and giving the advancing Allies something to worry about. The German Army might be in retreat *now*—that was obvious enough—but there would come a time when it would stand and fight, and make the Allies pay an impossible price for every yard of ground they gained. The last stand of the Thousand-Year Reich would be in the frozen wilderness of the Bavarian Alps, a splendid location for anyone devoted to the works of Wagner and the ideal of a heroic *Götterdämmerung*, but in military terms a natural fortress which could be held by comparatively few determined defenders.

Unfortunately such an eventuality seemed all too likely in the minds of many Allied intelligence experts. And, as happens so often when the enemy's actions and sayings tend to reinforce one's own picture of a situation, then that picture becomes more and more definite, until factors like the need for solid proof or the intrusion of common sense become irrelevant. The Allies were worried about the idea of a fanatic last-stand operation in the Bavarian mountains because, militarily, it was about the most difficult problem which the battered and virtually defeated German armies could still credibly present them with. It was completely in tune with German folk myths and Wagnerian drama so beloved by Adolf Hitler. And now here was Goebbels, not only chief of propaganda, but one of the top men in the Nazi hierarchy, saying that this was exactly what was going to happen!

It was hardly surprising that he was taken seriously. In September, 1944, three months after the invasion and before the German forces began to crumble, the American Office of Strategic Services, the OSS, predicted that Nazi Government departments would be evacuated to Bavaria as Germany was overrun. They started keeping a watch on the area, and soon reports began to come in from agents on the spot, from informers in neutral countries and through diplomatic channels.

So far, Allied Intelligence had been right to bear the possibility — even the probability — of a last-ditch stand by the Nazis in mind. And keeping a sharp look-out was a sensible precaution. But from here onwards the myth took on a life of its own, ignoring the facts in wild speculation and flights into mythology. On 12 February, 1945, with the Ardennes offensive over and the Allied armies approaching the Rhine, the United States War Department solemnly issued a counter-intelligence report announcing that Berchtesgaden would be the head-quarters of the National Redoubt.

Four days later, Allied agents in Switzerland passed on a report from neutral military attachés in Berlin which was much more serious. It stated the Nazis were 'undoubtedly preparing for a bitter fight from the mountain redoubt' and referred to 'strong-points connected by under-ground railroads'. It went on to say that 'several months' output of the best munitions had been reserved, and almost all of Germany's poison gas supplies. 'Everyone who participated in the construction of the secret installations', said the report, 'will be killed off . . . when the real fighting starts'.

This was enough to curdle the blood of the Allied commanders, who already had their hands full fighting an army which, even when tired, outnumbered and running short of weapons and ammunition, not to mention men, and fighting on totally unsuitable ground because of rigid orders from above, was still one of the most formidable opponents in the world. What might happen when the best-trained and best-equipped SS divisions withdrew into an impregnable mountain fastness with carefully built fortifications and stockpiles of supplies and ammunition? British Intelligence experts, together with their OSS colleagues, tried to damp down the increasing speculation by issuing more cautious appraisals of the possibilities, but they felt — rightly — that ignoring the threat altogether would be foolish.

On 11 March, 1945, SHAEF Intelligence warned Eisenhower that the Nazis were planning to build an impregnable fortress in the mountains, where Hitler himself would command the defences from his home at Berchtesgaden. These mountains were 'practically impenetrable', and, if that were not enough, 'Here, defended by nature and the most efficient secret weapons yet invented, the powers that have hitherto guided Germany will survive to reorganize her resurrection; here armaments will be manufactured in bombproof factories, food and equipment will be stored in vast underground caverns and a specially selected corps of young men will be trained in guerrilla warfare, so that a whole underground army can be fitted and directed to liberate Germany from the occupying forces.'

On 21 March, hours before Patton's Third Army, fresh from the conquest of the Saar, crossed the Rhine at Oppenheim near Mainz, Bradley's Twelfth Army Group Headquarters announced what it called a 'Re-orientation of Strategy'. Claiming that Allied objectives had changed, 'rendering obsolete the plans we brought with us over the beaches', it went on to propose that the idea of a thrust on Berlin, strongly urged by Montgomery, could no longer occupy 'a position of importance'. All indications were, according to the announcement, that 'the enemy's political and military directorate is already in the process of displacing to the Redoubt in Lower Bavaria'. General Bradley went on to propose that his army group split Germany into two halves by a strong drive through the centre, by-passing Berlin well to the south. This would not only stop German forces from the north of the country withdrawing into the redoubt, but would later allow the American forces to swing southwards to crush any remaining resistance in the mountains.

Four days later the Intelligence chief of the American Seventh Army in the south was predicting the creation of an élite force within the redoubt of between 200,000 and 300,000 men. Agents' reports had told of three to five long supply trains carrying new types of guns arriving each week since the beginning of February. Others spoke of the building of an underground factory, capable of producing Messerschmitt fighters.

By now the Allies were in a quandary of their own making. Major-General Kenneth Strong, SHAEF Intelligence Chief put the problem in a nutshell when he said to the Chief of Staff, General Bedell Smith, 'The redoubt may not be there, but we have to take steps to prevent it being there'. Bedell Smith, for his part, thought, 'There is every reason to believe that the Nazis intend to make their last stand among the crags'. And even Eisenhower's superior, General Marshall, referred to the formation of organized resistance areas. 'The mountainous country in the south is considered a possibility for one of these.'

So it was that the course of the Allied advance was changed. Eisenhower abandoned the headlong rush for Berlin in favour of a drive through central Germany to meet the Russians on the Elbe, to split the country in two and cut off all communications between the north and south. But when, on 16 April, Eisenhower turned the American Seventh Army south towards the mountains of the Bavarian Alps, they found next to no resistance at all, apart from the remnants of the German Fifteenth Army, who had come a long way from their secure but useless fortifications in the Pas de Calais. There were some Government leaders and groups of officials and some scattered SS units, but these had withdrawn from the north as the Russians and Americans had begun to close in. Apart from traffic jams with refugees and army units trying to surrender, there was nothing to stop the Americans from going wherever they wanted—certainly no underground fortifications, no miracle weapons, no poison-gas depots, no guns, no missiles and no jet fighters.

In the end the Ardennes offensive, the rocket attacks and the myth of the National Redoubt all ended in failure. Yet their final effects were to outlive Hitler and the Reich, firmly setting the shape of the cold-war battlefield of today. Because the Führer chose to expend his last reserves of troops and ideas in the west rather than in the east, the Russians were able to drive further and further westward.

First of all the Allies had to stop the Ardennes offensive before they could regroup, ready for their own push into Germany. Then the idea of the National Redoubt paralysed Allied thinking at the very moment when Eisenhower's armies were poised to strike for Berlin. The fears of a last desperate Nazi stand in southern Germany split their efforts. Although it was to add not a single hour to the life of the Nazi state, it gave the Russians a powerful post-war advantage. First into Berlin, they were able to lay their preparations for a complete takeover of the areas they had occupied in central and eastern Europe.

Alliances forged when self-preservation was the common aim were beginning to crack as the new priorities of peace pulled in different directions. Soviet Russia replaced the immediate aim of the defeat of Germany with her old objective of the spreading of Communism and the Soviet system throughout the world. Profoundly suspicious of Western motives, the Russians kept their forces on a full war footing. Two years after the German surrender they had 175 divisions fully operational and a force of 25,000 tanks, the majority of which were stationed in Europe, facing their 'allies'.

Thus was the scene set for the new 'peace'. The Russians greatly outnumbered the Western Allies in conventional forces in Europe, but they were outweighed in the final analysis by the atomic weapons which had forced Japan to surrender. Since neither side could afford war, the only alternative was to pursue one's objectives by every possible means short of actual fighting, and this threw psychological warfare into even sharper perspective.

Behind the crumbling façade of Allied unity, and the surface courtesies between the military governors, the Russians had begun a new and still more virulent propaganda offensive, directed this time towards the Germans living under Allied occupation. Fortunately this heavy and carefully planned psychological attack failed because it ignored the major requirements of all basic propaganda — that it must either be true or appear to be true. Instead Russian radio and newspaper reports claimed that Germans in the Western Zone were being coerced into forced labour on such a scale that they were fleeing to the Soviet Zone in ever-increasing numbers and causing food shortages. Not only was the reverse true at the time — that the harsh realities of Soviet rule were sending a flood of refugees westwards — but, more importantly, the Western Germans knew that this was the case from the evidence of their own eyes and ears. Soviet accusations that Marshall Aid from America was meant to impoverish the Germans made no sense to a people slowly recovering from near-starvation. And because of these obvious lies, the rest of the Soviet accusations — in particular that the Western Allies were preparing to abandon Berlin — were totally disbelieved by the population.

Yet, at the same time, whether they were believed or not, the

Russians still held the whip hand. On 15 June, 1948, they closed the autobahn bridge at Magdeburg for 'repairs', and with it the road route into Berlin from the west. Six days later water-borne traffic on the canals crossing the zone borders was stopped, and with it passenger traffic on the railways and all mail services. Three days later railway goods traffic was stopped, electricity supplies from power stations in the Russian part of the city were cut off and threats were made to throw the British out of a railway goods yard on their own side of the canal which formed part of the sector boundary. General Herbert called up a battalion of the Worcestershire Regiment, who were put to work obviously fortifying defence positions overlooking the canal bridges in the disrupted area. At the same time a troop of the 11th Hussars in armoured cars took up positions in the area, and the Royal Engineers prepared strings of mines which were held ready to close the streets and bridges if the Russians made a move.

The situation was tense, but Herbert's reasoning was sound. He knew the Russians on the spot might threaten and might take advantage of any slackness or gaps in the Western defences, but an attack on prepared defences, no matter how thin, was a very different matter. This would have to be a supreme-command decision from Moscow, taken only after very careful calculations of the results and consequences. On the other hand, he could ill afford to keep his troops, more than a third of his total force, waiting for an attack which would probably never be delivered. Once the Russians took up positions opposite him, with troops which could easily be spared from their vast resources, he would be a prisoner of his own actions; so, having forcibly made his point that Russian moves would be swiftly resisted, that night the troops were sent back to their barracks.

There was only one chink in the Russian blockade—the air corridors into Berlin. Since the Russians could scarcely interfere with this traffic without an act of war, this could still proceed almost unhindered. Indeed it could be stepped up, but the Allies doubted that they could do more than supply their own garrisons by air. What would happen to the vast civilian population of the Western Zones of Berlin? Yet herculean efforts were made to keep the city fed and supplied, and for eleven months from June, 1948, to May, 1949, the city was kept alive by air supplies. At the peak of the air-lift, eight thousand tons of supplies a day were being flown in. This was a big drop from the twelve thousand tons a day before the blockade, but at least there was enough to eat for soldiers and civilians alike. There was coal for the fires and factories, there was petrol for transport. Aircraft were landing every thirty seconds at peak hours and the food ration for the Germans actually went up during the blockade.

Psychologically, the air-lift inflicted a defeat on the Russians. They threatened to hold air exercises close to the corridors leading to Berlin, which might result in Allied aircraft being involved in accidents. But the threat was never carried out and British Intelligence was firm in the opinion that the Russians' innate caution would never allow them to embark on a course of action which would definitely result in war. The danger was in escalation through over-reaction, and the policy was to be firmness without provocation. Fighter escorts were avoided, and attempts to crash the road barriers with armoured columns were rejected. On the other hand, all measures on the Russian side which might make the air-lift more difficult or dangerous were to be resisted. Russian barrage balloons near their sector boundary made the approach to Gatow airfield difficult under certain conditions. A British official complaint resulted in the Russians lowering the balloons. On the other hand, the success of the air-lift was obtained at a price: sixty-eight airmen died in accidents before the blockade was lifted on 12 May, 1949.

The ultimate effect of the blockade, and the air-lift which resulted, was to bring about the speedy formation of NATO, which was another setback from the Russian viewpoint. Yet their short-term objective, the convincing of the German people that their only hope lay in accepting reunification of their country under Russian terms, was equally in ruins. Already those Germans living in the Western Zones of Occupation had good reason to know how fortunate they were. When General Sir Gerald Templer was appointed Director of Civil and Military Government in the British Zone in 1945, he had set in train a series of urgent measures which were vital to help the Germans survive the approaching winter. He put the British Army to work cutting down forests to provide firewood and pit-props to restart the Ruhr coal-mines; he drafted German PoWs to work on the land and down the mines, and he devoted the army engineers' efforts to restoring transport. In six months the Rhine was clear for barge traffic, and eight thousand miles of railway line were functioning once again. Plans were under way for pre-fabricated housing for more than a million Germans, and food imports were increasing. Books were being printed, newspapers and broadcasting were being restarted, and life was beginning, slowly, to return to normal. The flood of refugees from the east showed that things were not working so well on the other side.

Yet for all the tension across the European zonal frontiers, the first real post-war flashpoint occurred far away on the other side of the world. So effectively had the pressure been kept up on NATO that when the first all-out war since VJ-Day broke out in Korea, the West was taken entirely by surprise. There was one small consolation: the Russians, too, were taken unawares. Their absence from the Security Council of the United Nations also meant that their customary veto was missing. As the North Korean army poured south across the 38th Parallel, forty-one member nations of the UN voted for direct action and within ten days American troops were landing in South Korea, backed by air strikes to slow up the enemy advance.

But it was later, when the UN forces' advance northwards brought in Chinese 'volunteers' in their hundreds of thousands to stop the rout of the northern forces, that a new kind of psychology of war emerged. These new and unexpected opponents were a much more formidable foe altogether. Although often relying on head-on attacks quite literally regardless of cost, their guerrilla background made them past masters in the art of stealthy infiltration. Allied air superiority counted for little against an army which relied on mule transport, and which kept to unobtrusive tracks rather than the easily watched roads. Skilful use of camouflage kept British and American commanders in ignorance of their whereabouts or intentions, and attacks came with great skill and little warning. Often Chinese advance parties mingled with refugee columns to reconnoitre UN defence positions as they passed through. By New Year's Day, 1951, the troops were back south of the 38th Parallel, trying desperately to stem the Chinese tide, and still the attacks came. The Royal Ulster Rifles, dug in to the north of the capital, were confronted by men waving white flags and shouting that they wished to surrender. Seconds later, the full fury of the Chinese attack hit them in the pre-dawn blackness. By first light, the first they knew that one of the regiment's positions had been overrun was the sight of a Chinese bugler blowing reveille from the summit of the hill.

The Chinese advance was finally halted south of Seoul, and another advance northwards began, although this time it was slower and more careful as befitted the more formidable enemy. New ideas in aerial warfare were being brought into use to counter the Chinese skill in remaining under cover. Because they often moved only at night, special reconnaissance aircraft were sent up to keep a purely visual watch for movements. Any sign of movement at all was noted, and at first light reconnaissance aircraft would be sent in to check the area again before fighter-bombers struck with napalm, bombs and rockets. In time, the value of having air cover on tap, directed from slow-flying piston-engined light aircraft which could drop markers to guide the jet strike aircraft to their targets, was realized by the troops, and this too helped tilt the balance back in favour of the UN forces.

The Korean fighting now had more in common with the First War than the Second. Chinese attacks were preceded by artillery bombardments approaching those of the Western Front in intensity. But the no-man's-land was wide, and positions on the steep hillside were secure against sudden attack. The crying need of all the British commanders was for information, and the old First World War tactics of the trench raid to seize prisoners was brought into action. But the Chinese were very hard indeed to take prisoners. Many refused to surrender and died as a result, and even those captured because they were wounded invariably managed to die by the time they were brought back to the British lines.

Some units devised elaborate plans to lure the enemy into captivity. The Royal Norfolk Regiment deliberately left one hill unoccupied at night, moving their men into no-man's-land and waiting until the Chinese should take possession. Artillery fire plans were all laid down in advance, and for four nights they waited in vain. At last the Chinese reacted according to plan, and the attack was launched. Twenty Chinese were killed and six captured – but all died from their wounds, slight though some of them seemed to be, before they could be questioned. Finally, casualties from constant patrol activity forced many commanders to call a halt.

Other tactics were more successful. Attacking the Chinese at first light, supported by guns and tanks, sometimes caught them in their deep dug-outs, producing relatively cheap victories. But the fighting was always hard. In some of their counter-attacks, the Chinese themselves used shells by the thousand to soften up UN positions and demoralize the defenders. Often the attacking troops would come in through their own barrage for maximum shock effect.

Yet throughout the fighting morale was high. Units were never forgotten; visits by senior officers and entertainers were frequent, supplies of comforts were maintained once the war settled into its static phase, leave was regular and tours of duty were strictly limited. All was very different from trench conditions in the Great War, and the effect was obvious. The Commonwealth Division stayed in the line for more than eighteen months, far longer than the record for any equivalent unit under 1914–1918 conditions. And when an armistice was agreed and the shooting stopped, at ten in the evening on 27 July, 1953, the Chinese emerged at last to give the whole line a holiday atmosphere, with waving banners and shouted propaganda beamed from batteries of loudspeakers. Only the returning prisoners, with memories of hardship and privation, torture and ill-treatment, redressed the balance. Here the psychological attacks had been at their most brutal. Against lone prisoners, the most powerful inducements had been levelled to help persuade them to give information or to sign petitions supporting the Communist cause. But whereas the North Koreans had on many occasions tortured and even shot their prisoners, the Chinese methods were infinitely more subtle, rarely going beyond threats of shooting or other mental pressures to induce co-operation.

The standard process began with a friendly enough welcome, and a carefully framed appeal to co-operate in the signing of a simple peace petition, something which could be held to be reasonable whichever side a man supported. But this was merely the first step. Pressure was then increased. Questions were asked about the prisoner's home life and background to find out more about his indoctrination possibilities. Questionnaires of a detailed but non-military nature were issued under the stamp of the Chinese Red Cross 'to help in contacting the next-of-kin'. Lectures were organized to teach the prisoners the principles of Communism, on which they were asked questions and given snap tests at all hours of the day or night. Those who were unco-operative were singled out, and their whole group would often be punished for their mistakes or refusals. Even those who *did* co-operate were leaned upon heavily, and interrogations went on sometimes for weeks on end.

Routine was avoided—the unexpected was kept as a morale-breaking weapon as far as possible. Some prisoners would be kept waiting for hours on end for their turn to be interrogated, others would be questioned at short intervals at all hours of the day or night. Attempts were made to persuade all prisoners to write their autobiographies. Even where these contained nothing of military value, the background they revealed was made use of in later, more detailed, personal questioning as an aid to breaking down resistance. Letters from home were held up, only those with depressing contents being allowed through. Since these showed no evidence of censorship, many prisoners tended to assume the non-arrival of other mail was due to indifference at home or inefficiency in their own army, which made them feel still more isolated and vulnerable.

Some prisoners, inevitably, gave way under this psychological pounding. In most cases this took the form of signing peace petitions and other minimum forms of co-operation which have little or no real benefit to the other side. Incentives to yield included better accommodation, medical treatment, extra food and cigarettes. Others went further, and a very small group professed to embrace Communist ideas as a result of Chinese persuasion, although among the British prisoners this amounted to a minority of four per cent, most of whom had been confirmed Left-wingers before capture.

But the strongest lesson taught by those who survived Chinese captivity was that non-co-operation and group solidarity was, in the end, the best policy of defence. Those who resolutely refused even the slightest co-operation, either because of a deeply ingrained hostility to all authority, even their own officers, or because of very strong loyalty to their own training and upbringing, were usually given the hardest work to do. Often they were threatened or beaten to begin with, and given the severest handling by the interrogators, but once the Chinese realized their determination, they were left alone as hopeless cases. And wherever a group held together tightly, their treatment and chances of survival were greatly improved. Group solidarity among American prisoners tended to be relatively low, morale dropped severely, and the casualty rate was high: 38 per cent of American prisoners died in captivity. The Turks, on the other hand, stuck together and refused the slightest co-operation with their captors. Two hundred and twenty-nine were taken prisoner, some badly wounded, yet every one survived to the end.

In one respect at least Chinese psychological warfare had failed, especially in the strategic sense. For World opinion, outside the Communist bloc, adjudged them to be the aggressors in their attempt to drive the United Nations forces out of Korea.

Since then, confrontations of the scale of the Korean War have been rare and all-out war has been avoided. But psychological warfare goes on unabated. From both sides of the east-west border, propaganda crosses by rockets, balloons and radio waves from one half of Germany to the other. Perhaps the most dangerous is the brilliant black propaganda produced by the East Germans themselves. These include totally spurious but convincing-looking NATO secret documents, like those which prohibited American bombers from flying over United States territory while carrying live H-bombs—the implication being that bombs were habitually carried by American aircraft over other countries, like Germany. Others purported to be medical reports expressing great official concern over the increasing incidence of blackouts among American pilots.

Britain, too, came under attack. A forged British Cabinet document referred to a plot to limit the independence of trade unions in emergent African countries. Attacks were made on champions of European unity by the 'discovery' in the German archives of wartime documents discrediting their views. Other 'official' documents referred to plans to make West German conscientious objectors wear prominent badges, like those decreed for the Jews under Hitler.

No details were overlooked. When letters were sent to papers by the thousand to protest against West German rearmament, the East German writers took care that the names and addresses of the senders were genuine enough, except of course that the people named knew nothing of the letters sent in their name. Only detailed checking of all the correspondence isolated the forgers from the genuine protests. Other letters in feminine handwriting, hinting at clandestine affairs, were posted to the homes of soldiers and Government officers. And to make the leakage of official Western documents more convincing without the slur of espionage, the East Germans invented fictitious defections from West to East, which could account for floods of new, confidential information at regular intervals.

Other campaigns have been mounted to destroy morale and spread uncertainty within Western Germany: instructions on how to leave the army, sabotage equipment or routine, or avoid conscription were published by the West German Communist Party. False call-up papers have been sent to individual citizens; soldiers were sent false discharge papers; officials have been ordered to accept promotion or extra leave which did not in fact exist. All the headings and addresses are correct, and even when discovered to be forgeries, their convincing nature produces a growing sense of unease.

But apart from the purely psychological battle, both sides are continually trying to find out more and more about one another's defences, and it is here that the old wartime virtues of surprise and deception still play their full part. The need to know as much as possible about any potential opponent's reactions and capabilities is as pressing as ever, and this results in snooping and probing which goes on all over the world. No NATO naval exercise takes place without its uninvited guard of Soviet warships shadowing, listening, watching and noting every tactic, movement, signal or communication. For a long time, aircraft carriers in particular were an unknown factor to the land-oriented Soviet forces, and efforts to catch the smallest details of carrier-borne aircraft operations led to the collision between the *Ark Royal* and a Soviet destroyer which cut too close across the carrier's bows while manœuvring at speed.

14

Other confrontations occur regularly miles up in the stratosphere on the edge of British airspace. Soviet bombers, laden with electronic gear to pick up and record radar pulses and radio communications, approach the British defences to provoke them into reaction. Apart from the passive information picked up by their sensors on radar frequencies, there is more valuable information to be gained by what the defenders *do*. How long do the defending fighters take to arrive after the first emanations of the early warning radar had been picked up aboard the Russian aircraft? How long do they stay in company before having to refuel? What radar devices and weapons do the fighters carry? (Crews of both sides are briefed to take as many photographs as possible of the other aircraft.) What messages are passed backwards and forwards between the fighters and their ground control? What tactics and manœuvres do they carry out? What are their weaknesses?

This cat-and-mouse game creates the possibilities for all kinds of bluff and double bluff. Does a quick reaction to an approaching Soviet aircraft help to deter? Or does it give away too much? Is it best for fighters to stay in company to the last possible moment, giving away their endurance capability? Or is it best for them to be replaced by more fighters so early that they obviously have fuel in reserve, and nothing is revealed? At one time, Russian bombers when intercepted would descend to wave level, where the Lightnings used by the RAF interceptor squadrons would use up their fuel within minutes. So it became standard procedure for intercepting fighters to be supported by an airborne tanker a few miles away, enabling them to stay in contact for hours on end.

On land, too, the endless watchfulness goes on. Wireless traffic from army units on the other side of the border is picked up and analysed. Valuable information on tactics, operational procedure and even intentions can be deduced from monitoring these signals. For example, NATO listening units had several hours' warning of the Soviet Army move into Czechoslovakia in 1968 from the sudden change in the level of traffic over the Russian forces' radio networks.

In a sense, the endless confrontation and the ceaseless battle to ferret out one another's secrets works to the advantage of the Soviet bloc. For, given their huge preponderance in material, and their expansionist philosophy, much of the initiative remains firmly with them. By keeping such huge resources committed to Europe in the post-war years, the Soviets were able to keep Western attentions on the threat closest to home. This meant that aggressive preparations on the other side of the world came under less effective scrutiny, and guerrilla struggles were able to begin with all the advantages of surprise.

Even now, for all the emphasis on detente, some of the old realities of the Cold War confrontations still remain. The hard facts are that the preponderance in men and equipment of the Soviet bloc forces has grown rather than diminished with the years. In spite of hopeful developments like the talks on limitations of strategic weapons, and on mutual balanced force reductions, little has yet been achieved. Carefully worked out proposals for the kind of inspection of one another's forces and installations which alone could make these threats work have still to be made acceptable to both sides, and in the meantime commanders must gain the information they need on their opposition's capabilities and intentions as best they can. So instead of the old emphasis on brute force and bluster, today's undercover struggle has entered a new phase of guarded watchfulness.

The equipment for this long-range sentry duty has become capable of awe-inspiring achievements. Since the rejection by the Russians of President Eisenhower's 'Open Skies' mutual-inspection proposals in the 1950s, the Americans went to work on other methods of surveillance, principally the extraordinary U2 high-altitude, long-range spotter aircraft. This was a lightly loaded powered glider, which could fly at altitudes then far above the range of heavy interceptor fighters or even anti-aircraft missiles in the upper stratosphere. Using super-sensitive cameras, which could record details like golf balls on a green ten miles below, the U2s could cover forbidden areas of Russia in a depth never before possible.

Disguised as meteorological observation squadrons, the U2 units were based in Germany, in Turkey, in Pakistan and Japan. They carried equipment which recorded information on radar signals picked up by the aircraft, equipment which could analyse atmospheric samples to provide data on recent nuclear tests, and above all the high-resolution cameras. The information they brought back was invaluable. To take one example, Western observers at a 1955 Moscow flypast were horrified to see rank upon rank of Russian heavy bombers flying past overhead. Only a U2, miles overhead, was able to record that once out of sight, the leading Russian bombers were detouring back to rejoin the tail of the column for repeated passes over the watching crowds. The implication, that the Russians were content to place massive reliance on manned bombers, was shown to be false, and was confirmed by other U2 pictures showing the increase in intercontinental missile sites which showed the true story. Small wonder that the then CIA chief Allen Dulles, head of the U2 operations, said afterwards that the capabilities of the aircraft 'could be equalled only by the acquisition of technical documents directly from Soviet offices and laboratories'.

But the promise of the U2 flights crashed with the shooting down of Powers' U2 over Russian soil in May, 1960, and the collapse of the summit conference in Paris in the explosion of Soviet anger that followed. From then onwards, these remarkable aircraft would be restricted to more limited objectives, vital though these still were. It was a U2 which first alerted the Central Intelligence Agency to the arrival of Soviet ballistic rockets on Cuba in the summer of 1960. Following the confrontation between Kennedy and Khruschev which resulted, it was another U2 which confirmed that the Russian climb-down was genuine, and that the missile sites were indeed being dismantled.

Since then satellites have taken over where the specialized spotter planes left off. This time, their orbits can take them over opposing countries with impunity, and for the first time Intelligence chiefs like the director of the CIA (and his Soviet opposite number) have regular detailed coverage of the entire world. Apart from details revealed in a single shot, the fact that regular orbits cover the same ground time and again and can reveal the slightest changes in sharp relief adds a new dimension to Intelligence. To speed up the flow of information, some satellites drop capsules back to earth while others radio data from which pictures can be quickly built up on the ground. Newer projects are being developed which will provide continuous flows of information back to ground stations through other relay satellites. Whenever anything unusual emerges on the screens, ground controllers can then switch to extra high-resolution cameras on board the satellite for more information. For example, American Intelligence observation has been able to identify two different types of ballistic missile silos used by the Russians from a difference in diameter of only fifty inches or so.

In strategic terms, this gives the ability to see at last exactly what lies over the hill. But tactically, too, new aids to reconnaissance give commanders far more information than their predecessors enjoyed. High-speed low-level air reconnaissance can now reveal movement at night using super-sensitive film. Infra-red false colour photography makes camouflage virtually impossible since living foliage shows up a totally different colour from dead foliage or camouflage paint. Even where vehicles are hidden carefully under undamaged trees, the grass crushed by their tracks stands out clearly, showing their whereabouts.

Helicopters, because of their ability to hover, can carry even more sensitive sensors, like the American-developed 'people sniffers' used in Vietnam, which detected chemicals present in human perspiration.

Combat aircraft, too, grow ever more sophisticated. Bombers have long been capable of launching flares which attract heat-seeking missiles away from the aircraft itself, and pilotless drones which provide a stronger radar echo to attract electronically guided weapons. Now the aircraft themselves are taking the offensive against the hitherto invulnerable defenders. American Phantoms carry a system called Wild Weasel which picks up enemy radar signals and, at the appropriate moment, launches a missile of its own which flies down the radar beam to knock out the defences.

In one sense, deception and misleading the enemy becomes ever harder. At the end of the last Arab–Israeli War, so confused was the military situation that in many cases the commanders on the spot did not know the whereabouts of all their own units, let alone those of the enemy. The only people who *did* know the exact dispositions of both sides were the American and Russian Intelligence chiefs, from their satellite observations. Yet deceptions are still possible even in the 1970s, against all the technology of a modern army. In Vietnam, for example, the Americans used high-altitude reconnaissance to keep watch on North Vietnamese progress in repairing vital bridges destroyed by American attacks. Only when repairs were virtually complete would they be attacked again, cutting American losses and tying up North Vietnamese resources to the best advantage. But the North Vietnamese soon developed the technique of building replacement spans just under the surface of the muddy rivers, where they were hidden from the cameras, but which still allowed traffic to cross whenever the planes were not in the area.

For this reason, the North Vietnamese were quick to demand the ending of reconnaissance flights as condition for opening negotiations towards a ceasefire. Yet their own experience shows that while deception and surprise become more difficult to achieve in full-scale war, the value of either has never been higher. Simply because armies, and even nations, now know so much about one another's movements and intentions—from the footfalls of a single sentry to the advance of a squadron of tanks, from gaps in the scan pattern of an air-defence radar to the building of a missile platform—then the opportunities for successful deception, or indeed for the fruitful use of any kind of tactic of psychological warfare, begin to bristle with new difficulties.

Yet there is a paradox here. As the difficulties of bringing psychology to bear on one's opponent multiply, simply because of the vastly greater weight of evidence which must be provided to satisfy all this greatly increased and considerably more effective surveillance, then this greater weight of evidence must itself make the deception more credible. In other words, if a deception *can* be mounted which satisfied all the observers, then it is far less likely to be disbelieved than were many of the simpler schemes of the past. In the shadowy, undercover world of tomorrow's guerrilla confrontations, as in the worldwide watchfulness needed to maintain the careful balance of detente, the age-old weapons of psychological warfare have surely never been more powerful, nor more vital.

INDEX

Abwehr (German Military Counter-Intelligence), 102, 108, 112, 117, 119, 123, 136
Aden, 173
Admiralty, 8, 49, 85
Afrika Korps—*see* German Army
Agents (see also under codenames of individual agents—Garbo, Brutus, etc), 110, 112, 113, 114, 116, 118, 119, 123, 124, 129, 135, 136, 154, 155, 180, 181, 182
Airborne Division, 82nd (US), 124, 125
Airlift, Berlin—*see* Berlin Airlift
Alamein, 35, 36, 37, 102
Alam Halfa, 36, 37
Alexander, General Sir Harold, 106, 107
Alexandria, 60
Alfieri, Dino (Italian Ambassador to Berlin), 79, 80
Algeria, Allied landings in, 104
Ambushes, Malayan campaign, 160
Americans, 80, 102, 104, 105, 116–118, 128, 150, 153, 155, 179, 199
Amiens, 14, 15, 16
Andenne (Belgium): German terror-tactics in, 2
Aphrodite (radar-reflector on U-boats), 48, 49
Arabs, 169, 170, 199
Ardennes, 27, 182, 184, 187
Arras, 9
Associated Press (leak of D-Day news), 137
Athenia, sinking of, 41, 42
Atlantiksender (*Deutsche Kurzwellensender Atlantik*—British 'black radio' station), 81–84, 86
Australians, 16, 35, 39
Austria, 22

BBC broadcasts (*see also under* 'Radio'),
30, 60, 75, 87, 95, 136, 137
B-Dienst (German codebreaking service), 42, 43
Beaverbrook, Lord; and World War I propaganda, 17
Bedell Smith, General Walter, 187
Beersheba, deceptions at (World War I), 11–13
Belgium, 1–3, 26, 27, 70, 108
Berlin, 56, 186, 188
Berlin airlift, 189, 190
Berlin, Russian blockade of, 189
Bevan, Colonel (CO of London Controlling Section of Allied deception unit), 114, 119, 123
Biscay, Bay of, 44, 48
Black Boomerang (by Sefton Delmer), 76
Black propaganda—*see under* 'propaganda, black'
Black radio—*see under* 'radio'
Bleicher, Hugo (Abwehr man in Paris), 117
Bodenschatz, Gen Karl (Goering's aide), 54
Bomber Command—*see under* Royal Air Force
Borneo, confrontation campaign in, 167, 168
Braddocks (incendiary devices), 100, 101
Bradley, General Omar, 135, 149, 186
Breslau, 6
Briggs, Sir Harold (Director of Operations in Malayan campaign), 161, 162
Briggs Plan, 161, 162
British Army, 3–5, 9, 10, 13–16, 27, 28, 150, 157–199
    Units: 1st Armoured Division, 27
        Guards Armoured Division, 150
        3rd Infantry Division, 150
British Intelligence, 25, 49, 58, 106, 114, 119